THE RED RIVER BRIDGE WAR

Red River Books, sponsored by
Texas A&M University–Texarkana
Doris Davis, General Editor

THE RED RIVER BRIDGE WAR

A Texas-Oklahoma Border Battle

RUSTY WILLIAMS

TEXAS A&M UNIVERSITY PRESS
COLLEGE STATION

First edition

This paper meets the requirements of
ANSI/NISO Z39.48–1992
(Permanence of Paper).
Binding materials have been chosen for durability.
Manufactured in the United States of America

Library of Congress Cataloging-in-Publication Data

Names: Williams, Rusty, 1948– author.
Title: The Red River Bridge War : a Texas-Oklahoma border battle / Rusty Williams.
Other titles: Red River Valley books ; no. 4.
Description: First edition. | College Station : Texas A&M University Press, [2016] | Series: Red River Valley books ; number four | Includes bibliographical references and index.
Identifiers: LCCN 2016004932| ISBN 9781623494056 (book/cloth : alk. paper) | ISBN 9781623494063 (e-book)
Subjects: LCSH: Red River Bridge Controversy, 1931. | Texas—History, Military—20th century. | Oklahoma—History, Military—20th century. | Texas—Economic conditions—20th century. | Oklahoma—Economic conditions—20th century. | Depressions—1929—Texas. | Depressions—1929—Oklahoma. | Transportation and state—Texas. | Transportation and state—Oklahoma. | Toll bridges—Red River (Tex.-La.)
Classification: LCC F391 .W695 2016 | DDC 976.4/062—dc23 LC record available at http://lccn.loc.gov/2016004932

CONTENTS

THE RED RIVER BRIDGE WAR

INTRODUCTION

Coarse-Looking Men

FIFTY-THREE-YEAR-OLD Charley T. Davis was the first to lose his job as a result of the Red River Bridge War.

The work wasn't much, but Davis certainly needed the paycheck. Northern Texas and southern Oklahoma were feeling the pinch of financial depression. Times were hard, money was tight, families were frightened, and jobs were scarce. Davis was the sole support of his wife and two unmarried daughters. He had been a traveling salesman, a barber, and a grocer, but now he spent ten hours a day sitting in a tiny watchman's hut, earning $2.30 a day in state pay for guarding a highway bridge that had yet to open for traffic. It was 1931, and he was glad to have the work.[1]

Davis guarded the Texas entrance to a brand-new concrete-and-steel bridge constructed over the Red River to connect the towns of Denison, Texas, and Durant, Oklahoma. It was a final link in a chain of state highways and bridges stretching from the ports at Galveston, Texas, to the Canadian border. Completed just three weeks earlier, the new bridge was intended to replace a rickety old private toll bridge whose owners squeezed every farmer, dairyman, factory worker, and tourist for seventy-five cents to travel the thousand feet across the river that separated Oklahoma and Texas. Committees from the two nearest cities planned an all-day celebration to mark the opening of the free bridge on July 1, 1931.

The toll bridge owners, however, had other plans. They petitioned a federal court to grant them a restraining order against Texas, contending that they had exclusive rights to a bridge at that crossing and Texas was refusing to compensate them for those rights. When a judge granted an injunction prohibiting Texas from opening the new bridge, Gov. Ross S. Sterling of Texas, a tight-vested law-and-order man, ordered that the new bridge remain closed until the legal mess could be sorted out.

Charley Davis was a local fellow and a friend of the Grayson County sheriff. The sheriff hired him as a private watchman to patrol the Texas side of the free bridge, posting him there on Governor Sterling's orders to be sure the injunction was honored. Davis, unarmed, had little to do but sit in his tiny guard shack near the flimsy wooden barricade that obstructed the southern bridge entrance, occasionally directing an errant driver downriver toward the toll bridge. The work was uneventful and the pay was fair.

By Thursday, July 16, 1931, the gleaming new bridge had been standing idle for almost three weeks, and no end to the legal logjam was on the horizon. At 5 p.m. that day, Davis watched a dark sedan drive onto the northern approach road and park next to the sawhorses and oil lamps blocking the Oklahoma end of the bridge. Technically, Davis was responsible only for the Texas side, but he put on his spectacles and squinted a little to watch four men exit the car and begin to move pieces of the Oklahoma barricade to the sides of the road. With the north entrance clear, the four men returned to their vehicle and drove slowly south across the bridge toward Davis. The sedan stopped ten yards short of the guard shack; the four men once again got out, this time armed with tire irons, pry bars, and an axe.[2]

"They were coarse-looking men," a shaken Davis would say later. "I was completely unaware of their purpose." The men walked to where the watchman stood, surrounding him—one in front of him, one behind him, and one each to his left and right. The man in front reached into his pocket, slowly withdrew a folded letter, and handed it to Davis.

"I'm Burt McCaleb with the Oklahoma Highway Department," the leader of the group said when they reached Davis. "We're here under the orders of Governor William H. Murray to open our bridge." McCaleb gave Charley Davis a copy of Murray's letter, telling the watchman he had orders from the governor to "take down" the old watchman if necessary.

A stunned Davis retreated from his post, leaving McCaleb's Oklahoma crew to walk into Texas with their pry bars and tire irons and destroy the wooden barriers blocking the south end of the bridge. By the time Davis reached a telephone and reported to his bosses, Texans and Oklahomans were zipping across their new free bridge at the rate of seven hundred cars an hour.[3]

Later that evening, Governor Sterling dispatched a contingent of Texas Rangers with orders to close the new bridge immediately and keep it closed. Upping the ante, Governor Murray sent crews to tear up the highway approaches to the toll bridge. He promised to send armed Oklahoma National Guardsmen to the bridge if his men were interfered with.[4]

Charley Davis was out of a job. Oklahoma and Texas were at war.

At the beginning of America's Great Depression, Texas and Oklahoma armed up and went to war over a seventy-five-cent toll bridge that connected their states across the Red River. For two weeks during the sweltering summer of 1931—as traffic backed up for miles, commerce ground to a halt, and the Great Depression spread—Texas and Oklahoma waged a legal and territorial conflict marked by finger-pointing, regional bombast, and the outrageous posturing of two of the most willful state governors of the twentieth century.

Today, more than eighty years later, the Red River Bridge War has passed almost entirely from living memory into the realm of folklore, yet the full story of the conflict has never been told. Most accounts paint the two-week standoff at the bridges as a unique eccentricity of the times, something on par with flagpole sitters, marathon dancers, and wing walkers. It is little wonder that Oklahomans who have heard of the war (and many of them have) think it to be a case of Texas arrogance, resolved only when the Oklahoma military sent a squad of Texas Rangers running by retaking and reopening a highway bridge across the Red River. Texans who have heard of the war (and few of them have) think it to be a case of lawless Oklahomans, spurred by their rural bumpkin of a governor, who defied court orders and appropriated a highway bridge both states had paid for.

Neither version is correct.

The Red River Bridge War was never a full-on shooting war (although a few of the Texas Rangers did some target shooting down in the river bottoms). To those not directly affected, it was a comic-opera war marked by quilting bees on the barricaded span, midnight bulldozer raids, a peace delegation of Native Americans, secret treaties with Spain, and blockade runners in Model T jalopies. Newspapers shared the comedy with readers across the nation with banner headlines. The *New York Times* and the *Washington Post* posted correspondents at the scene; *Time* printed tongue-in-cheek profiles of the combatants.

But the Red River Bridge War was a serious clash between two eccentric state governors, each with his eye on national office. Ross S. "Fat Boy" Sterling was the archetypal Texas oilman-turned-politician. Founder of Humble Oil (the predecessor of Exxon), owner of Houston's two newspapers, and chairman of the Texas Highway Commission, he won election on a strict law-and-order platform and a promise to impose his executive competence on the statehouse. William H. "Alfalfa Bill" Murray was a pistol-packing homespun populist, a rabid agrarian who lived off the proceeds of his own small family farm. Active in Oklahoma politics even before statehood, he came to the state capitol in 1931 with a mandate to protect small farmers from foreclosure and bank failure. As different in their outlooks as presidents Hoover and Roosevelt, the two governors acted out their conflicting philosophies of government at a time of plummeting tax revenues and financial uncertainty.

Neither man was particularly forthcoming about his role in the bridge war at the time, nor in later years. Governor Sterling never mentions the conflict in his autobiography or oral history. Governor Murray wrote his account of the bridge war some two decades after the event, and when the legend contradicted the facts, Murray recounted the legend.

National media attention shone a bright light on Sterling and Murray, and press coverage of the bridge confrontation helped propel one toward the White House and the other into political obscurity.

For ordinary city and farm families living on both sides of the river, the story of the Red River Bridge War is the story of the fragility of hope—hopes raised, hopes crushed, and hopes (eventually) restored—played out under gathering thunderclouds of the Great Depression.

It may be difficult today to appreciate how important a toll-free bridge crossing was to the largely rural population of more than eighty years ago. With the coming of the automobile at the turn of the twentieth century, the states (with some help from the federal government) assumed financial responsibility for the highways connecting their cities and rural communities. By 1930, however, despite three frantic decades of grading, graveling, asphalting, and concreting free pathways for an ever-increasing number of automobiles and trucks, America's farm-to-market road systems were still primitive at best. Meanwhile, the Depression was dealing a double whammy to rural populations, driving down farm prices while drying up local tax revenues for continued infrastructure investment. Free and reliable transportation routes from farm to market and state to state would be key to recovery from financial hardship for many Americans, and the standoff at the Red River bridge demonstrated how tenuous that grip on recovery might be. For all the national coverage, most journalists failed to report on the slow strangulation of optimism among local farmers, merchants, dairymen, factory workers, and travelers each day the new free bridge remained closed. The outlandish aspects of the Red River Bridge War drew attention from the increasing likelihood of bankruptcies, defaults, shortages, and food riots in the region.

The events at the two Red River bridges in 1931 also provide an object lesson for citizens of the twenty-first century, as modern political leaders choose between public works and privatization of highway infrastructure. Today, cash-poor state and local governments are again granting long-term concessions to private entities for the operation of toll roads and bridges, even as a simmering national financial crisis may require free and reliable transportation routes to speed recovery. Since 2004, thirty-one states have signed long-term agreements with private companies to convert thousands of miles of state highways, bridges, or routes to private tollways. It's good business for the private companies, but maybe not so good for the communities involved. The story of the Red River Bridge War is a cautionary tale, providing historical context to the current trend of reprivatizing our nation's highway infrastructure.[5]

Above all, the story of the Red River Bridge War is a very human tale of well-meaning individuals who did their best as they saw how to do it at the time. It includes the stories of an entrepreneurial Indian whose hard work and foresight make him one of the wealthiest men in Indian Territory; of an otherwise honest politician who trips over one little shortcut and falls into a pit of deception; of the son of a larger-than-life politician who is driven to win his father's final legal case; of a small-town state legislator who wants nothing more than to please the people who voted for him; and of a young newspaper reporter who has to tell his bride he's cutting their honeymoon short for the biggest story of his career. These and other colorful individuals played a role in bringing Texas and Oklahoma to arms (and to armistice) during that sweltering Depression summer of 1931.

Oklahoma's governor triggered the Red River Bridge War when he sent Burt McCaleb's highway crew into Texas to destroy the bridge barricades. But the wellspring of the conflict can be found more than a century earlier, when John Quincy Adams was forced to clean up a mess left by Thomas Jefferson.

CHAPTER ONE

Great Sagacity and Tact

BY MOST ACCOUNTS, Thomas Jefferson was something of a genius—a big-picture guy whose vision occasionally outstripped his ability to nail down every detail.

President Jefferson certainly recognized the big picture in 1803, when France offered to sell 828,800 square miles of North America to the United States. Jefferson understood that this land of "Louisiana" included the entire western drainage area of the Mississippi River, all the way to the Rocky Mountains. The Louisiana Purchase—a bargain at $15 million—would acquire river access to the Gulf of Mexico, double the size of the United States, and open American expansion to the north and west while blocking Spain's North American expansion plans. Despite some domestic opposition, President Jefferson and the United States marked the formal transfer of territory from France in Saint Louis on March 10, 1804.

The precise boundaries of the Louisiana Purchase were problematic, however. The seller didn't know precisely what he had to sell, and the buyer didn't know the exact extent of his purchase.

Jefferson dispatched Lewis and Clark and their Corps of Discovery to explore the northern and western reaches of the new lands, but the southern border—the boundary shared with an increasingly belligerent Spain—remained a matter of hot dispute. At one point Jefferson claimed that the Rio Grande should mark the border between the two countries; Spain placed the boundary 500 miles to the north, along the Rio Rojo de Natchitoches. For fifteen years, neither country pressed its claim to the land known as Tejas, but by 1819 the United States and Spain figured it was time to straighten things out.[1]

That's when John Quincy Adams went to work.

Son of founding father and second president of the United States, John Adams, John Quincy Adams was already an experienced diplomat when Pres. James Monroe appointed him secretary of state in 1817. Spain's influence in North America was waning: its colony in Mexico was in rebellion, and revolutionary activity in its Central and South American colonies left the European power with little military or government presence in Florida. In 1818 Secretary of State Adams sat down at the treaty table with Spanish Foreign Minister Luis de Onís to negotiate the acquisition of Florida and—while they were at it—settle the longstanding Tejas boundary dispute left over from the Jefferson administration years before.[2]

With the Adams-Onís Treaty of 1819, Spain turned Florida and coastal Louisiana over to the United States; the United States essentially renounced any claim to the Spanish colony of Tejas by agreeing to an international border of the Sabine River and the Rio Rojo. Adams and Onís also agreed to an odd provision in the language of the treaty (and their intent was discussed and questioned in American courts well into the twentieth century): they placed the border line between their two nations on the southern riverbank of the Rio Rojo, not midstream as was usual.[3]

According to the pact hammered out between the two diplomats and ratified by the US Congress in 1821, Spain (and its eventual heirs) could lay claim only up to the solid ground on the southern banks of the Rio Rojo, a rust-colored stream that Anglos would come to know as the Red River.

The Red River has its headwaters in the redbed lands of the Llano Estacado and the high plains of modern-day New Mexico. Water from snowmelt, fierce storms, and the occasional spring flows down the tributaries to the main river, carrying with it an iron-rich silt and sand already oxidized to the color of terra-cotta.

For much of its 1,300-mile length, the river flows east-southeast until it reaches modern-day Arkansas and swerves southward to meet up with the Mississippi River. There is not a mile of this upper river straight to the compass; along its course it doubles and twists back on itself like a rust-colored rattlesnake. And, like a rattlesnake, the Red River can spring from lazy coil to lethal strike in a blink. Mostly, the river is a slow-moving and shallow stream with rivulets wandering along the red clay streambed among sandbars, patches of quicksand, and clumps of opportunistic cottonwood trees. At other times—often without warning—the river can become a mile-wide torrent, scouring its banks, cutting new channels, and uprooting any natural or man-made objects in its path.[4]

The prairie lands south of the upper section of the Red River were once the shoreline of an ancient ocean, a littoral zone where rock was long ago beaten into rich, black earth. As the Red River alternately oozes and gushes, it mixes its iron-rich red sediment into the black earth, creating a purplish-black loamy soil perfect for growing cotton, wheat, rye, and corn.[5]

Farmers would see the promise of that land.

More interested in gold and souls than farms, the Spanish never had a chance to cultivate the Red River valley; the ink had barely dried on the Adams-Onís treaty when Spain lost Tejas to the new Republic of Mexico. Illegal infiltration into Tejas by Anglo-American settlers led to the colony's secession from Mexico and the establishment of an independent Republic of Texas in 1836. One of the first acts of the fledgling republic was to affirm all provisions of the Adams-Onís Treaty, adopting the Red River as the international boundary between Texas and the United States.[6]

The Republic of Texas lasted only nine years until, nearly bankrupt, it entered into a treaty of annexation with its larger neighbor. In 1845 Texas became the twenty-eighth state of the United States of America.

With the annexation of Texas, the Red River no longer marked an international boundary. However, the river still defined the border between Texas and a territory so alien that it might as well have been a foreign country.

For most of its history, up through the nineteenth century, the US government applied what came to be known as the "Three R" policy—removal, resettlement, and reservation—toward Native Americans. Indigenous peoples whose ancestral lands were in the path of settlement by European Americans (and expansion by the United States) would be removed and resettled in lands farther west. A generation later, as white settlers encroached on the new lands, the tribes would again be uprooted and resettled.

Five tribal groups that lived in areas south of Virginia after the American Revolution were repeatedly relocated from Georgia to Tennessee to Alabama to Mississippi and Arkansas between 1780 and the 1830s. Known as the Five Civilized Tribes, the Chickasaw, Choctaw, Cherokee, Creek, and Seminole were thought to be sufficiently advanced in the civilization of their European conquerors that they might be trusted to govern their own independent nation-states if lands were reserved for their exclusive and lasting ownership.

US government officials and tribal leaders began to accept the need for a permanent reservation, and in 1832 Pres. Andrew Jackson appointed a commission to explore the suitability of the territory west of Arkansas and north of the Red River. One by one the tribes voted, treaties were signed, and the Five Civilized Tribes relocated to the newly designated Indian Territory west of Arkansas. The Cherokees claimed the northeast corner of the Territory (along with a seventy-mile-wide strip along the Kansas border, giving them free access to the Great Plains). The Creek and Seminole received a drumstick-shaped slice in the middle of the Territory (but with fine farming land along the Canadian River). Chickasaw and Choctaw shared the bottom third of Indian Territory, from the Kiamichi Mountains and Arkansas in the east to the western reaches of the Territory. Their southern border was the Red River.[7]

The Five Civilized Tribes were not the war bonnet and breechclout Indians of later dime novels and movies. They had spent more than a century in peaceful proximity to the European settlers and their mission schools, and during that time adopted some important traits of European culture. The tribes had understood and employed improved transportation methods, domestic animals, and new plants—all useful for stabilizing food supplies and making markets. Four of the five tribes operated under written constitutions that spelled out elected representation, taxation, property rights, and lawful personal conduct; three of the tribes organized finishing schools and special academies for tribal members.

The Chickasaws, in particular, absorbed the lessons of capitalism from their alien invaders. The new Chickasaw Nation in Indian Territory was dotted with mills, trading houses, gins, and other businesses. Some Chickasaw merchants drove pack trains of goods westward to trade with the wild tribes; other Chickasaw entrepreneurs established spas and tourist facilities near secluded oil seeps and springs in the Arbuckle Mountains. A few tribal members built and operated ferries to help travelers and their wagons ford the rivers and streams that ran across Indian Territory.[8]

From 1850 to 1860 trails and roads leading westward from the Ohio Valley states teemed with settlers eager to claim rich Texas farmland south of the Red River. At a time when the total population of the United States increased by a third (including the addition of four new states), Texas' population tripled. The former republic received almost half a million immigrants in the decade of the 1850s, the equivalent of a small village entering the state every week of every year for ten years.

The route of choice for many of these hopeful immigrants was the Texas Road, a settlers' wagon route that crossed Indian Territory from its northeast corner near Joplin, Missouri, southwestward to the Red River. Hundreds (and later thousands) of white families traveled the Texas Road weekly, crossing tribal lands and paying fees for the privilege. Tribes established trading posts to resupply the travelers, charged tolls on improved sections of the road, and issued licenses to ferrymen who helped travelers cross the streams and rivers.[9]

After a trip of 260 miles through Indian Territory, travelers came at last to the crossing into Texas: Colbert's Ferry across the Red River.

"Mr. Colbert, the owner of the station and of the ferry, is a half-breed Indian of great sagacity and business tact," observed one of the travelers. "He is a young man, not quite thirty, I should judge. We crossed the wide, shallow and muddy Red River on one of Mr. Colbert's boats."[10]

Half-breed or not, Benjamin Franklin "Frank" Colbert bore a family name that was celebrated in the Chickasaw Nation. European traders and trappers crisscrossed America's southeastern territories in the eighteenth century, often marrying (and being adopted) into the Native American tribes they encountered there. James Logan Colbert married a Chickasaw woman in about 1745 and raised their sons as English-speaking tribal members. Son Levi recruited and trained a division of Chickasaw militia that fought alongside Gen. Andrew Jackson in 1816 at the Battle of New Orleans. Later, as the United States formulated plans to remove the Chickasaw from their ancestral lands, Levi Colbert traveled to Washington to negotiate directly with President Jackson. (Though the tribe had no such formal title, white writers referred to Levi Colbert as "Chief of the Chickasaw Nation.")[11]

A grandson of Levi, Frank Colbert was born in Mississippi in 1826. As a young man he worked for his uncles in the family ferry business before the removal to Indian Territory in 1846. After the move west, much of the Colbert family settled on rich cotton and peanut farming land on the north banks of the Red River. Frank went into the ferry business just as immigrants began to surge down the Texas Road. He was a hard worker with a compact build, barely over five feet tall. Thick muscles draped his arms and shoulders like a robe, and he had the square head and high cheekbones of his aboriginal ancestors. Frank Colbert was a natural businessman, but he was fully capable of the physical labor required for transporting people, loaded vehicles, and balky livestock across a half mile of moving water.[12]

His boat was "simply sort of a raft, pushed across the stream by the aid of poles," according to one early traveler. "The fare for a four-horse team is a dollar and a quarter, and the net revenue of the ferry about $1,000 per annum."[13]

There were other Red River crossings, but Colbert's Ferry was nearest the terminus of the Texas Road. On the Texas side, an improved road led from the ferry landing ten miles due south to the growing town of Sherman, Texas. From there, immigrants could continue sixty miles south to the trading post at Dallas or strike westward for the trip to Southern California.

But the greatest advantage of crossing at Colbert's Ferry was young Frank's doggedness in employing his "sagacity and business tact" to make the crossing as easy and pleasant as possible for his customers. His first ferry was a tippy wooden raft barely large enough to accommodate a single wagon and the men who poled the boat through the river's current. Quickly, Colbert built larger flatboats, some up to forty feet long and fifteen feet wide, capable of carrying a fully loaded freight wagon and its harnessed team. He replaced his pole teams with a cable and winch device that made the crossing faster and more stable. From time to time, floods would destroy his landings and wash the ferryboats downstream, but Colbert always had replacement boats and cable waiting in his workshop, ready to resume service as soon as the waters receded. Colbert hired engineers (and bought slaves) to improve the gradient and condition of roads leading to the landings.[14]

By the mid-1870s, Colbert was ferrying as many as two hundred wagons a day across the Red River.[15]

To protect his investment from interlopers, Colbert sought an exclusive franchise from the Chickasaw Nation. The tribal legislature granted him a license, requiring Colbert to "keep at all times good boats and trusty and efficient boatmen," keep the roads leading to the ferry in good order, and post his rates in a conspicuous place. The franchise protected him from competition for several miles up and down the river. It was valid until such time as the tribal legislature might act to revoke it.[16]

In 1858 the Butterfield Overland Mail instituted coach service to carry passengers and mail from Saint Louis to the California coast, and Frank Colbert saw an additional business opportunity. The Butterfield route entered Indian Territory near Fort Smith, Arkansas, and generally followed the Texas Road south to the Red River. Colbert's landing was designated an official stage stop, and his ferry would carry the stagecoach into Texas.

"Mr. Colbert evinces some enterprise in carrying the stage across his ferry free of charge in consideration of the increased travel it will bring his way," a newspaperman reported. The loss of a ferry charge for the stagecoach was nothing compared to the money made by feeding stagecoach passengers, maintaining teams of relay horses, and providing blacksmith services.[17]

In addition to his ferry business, Colbert operated one of the busiest stops on the Butterfield line. He provided lodging and meals to travelers, ran gristmills and a sawmill, had more than five hundred acres under cultivation, and owned twenty-five slaves. His employees and extended family owned cabins nearby, forming a thriving community that bore Colbert's name. By 1860 the thirty-five-year-old Chickasaw entrepreneur was considered to be one of the richest men in the Indian Territory.[18]

For four years the Civil War slowed ferry traffic, but Colbert's Ferry remained the crossing of choice for the wagonloads of agricultural and manufactured goods transported between Texas and the Indian Territory. By 1870, however, engineers for the railroad companies were beginning to scout sites for possible bridge crossings over the Red River.

The transcontinental rail link was completed from the East Coast to the West in 1869, and rail companies turned their attention to a north-south line connecting the Great Lakes to the port of Galveston, Texas. As the Houston & Texas Central Railroad laid track northward from Galveston, the Missouri, Kansas & Texas Railroad secured land grants from the tribes of Indian Territory to build southward from Kansas. The companies expected to meet up somewhere south of the Red River near Colbert's Ferry.[19]

When railroad bridge engineers arrived at the Red River to survey possible crossings, Frank Colbert welcomed them, fed them, and allowed them to camp on his land. He helped the engineers scout the riverbanks while he listened and learned about the particular difficulties of bridging the Red River. Colbert had certainly recognized that rail service would choke off much of his ferry crossing business from long-haul freighters; there was sufficient short-haul business, but he didn't want to lose that to the convenience and speed of a rail bridge crossing. In the end, Colbert found an appropriate bridge site for the railroad, a few miles upriver from his ferry. At the same time, he identified a site for a bridge of his own.

When the MKT ("Katy") Railroad rolled across its Red River bridge into the new town of Denison, Texas, on Christmas Day 1872, Colbert was well under way in planning for his own bridge.[20]

"He and I had several conversations in regard to it that spring," an employee recalled. "He went to Washington to see about getting a charter."[21]

Colbert was about to invest a goodly portion of his fortune in an expensive wagon bridge. He asked the Chickasaw legislature to renew his franchise, adding operation of a toll bridge. He also wanted to be sure the US government would give him at least the same protections from competition that his tribal government provided.

A Chickasaw resident of Indian Territory, Colbert had no official representation in Congress, but he was prepared to dance the capitol quadrille like a professional. He carried with him a letter from the governor of the Chickasaw Nation describing how a wagon bridge across the Red River would serve the interests of the Five Civilized Tribes and petitioning the Commissioner of Indian Affairs for the protection by the US government of such a bridge. In a matter of seven days in Washington, Colbert walked Governor Harris's letter up through the offices of the commissioner of Indian affairs, the secretary of the interior, the speaker of the House of Representatives, and the president of the United States.[22]

"When he came back he told me that he got an introduction and shook hands with the president, and he was surely proud of it," the employee recounted. "I asked him if he would not have to get some authority from Texas. 'No,' he said, 'the Chickasaws claim to the high water mark on the south side of Red River.'"[23]

Perhaps he had second thoughts, because shortly after his return from Washington Colbert traveled to the Texas capital of Austin to ask the state legislature for a franchise from that state. Texas had every reason to foster transportation across the Red River and accommodated Colbert, passing a bill granting the Chickasaw ferry and bridge exclusivity for three miles up and down the river. One provision of the law, however, limited Colbert's franchise to just twenty-five years.[24]

On February 2, 1873, Rep. John C. Conner of Texas and Rep. Isaac C. Parker, freshman representative from Missouri, introduced HR3917, a bill drafted almost entirely by Colbert and the Chickasaw governor. Noting that US government approval probably wasn't necessary, the bill nevertheless called for congressional authorization for "B. F. Colbert, a Chickasaw Indian, his heirs and assigns, to build a transit and wagon bridge across the Red River." The bill required that Colbert allow US postal carriers to cross the bridge at no charge and that telegraph lines be strung over and across the bridge. The law also made it unlawful "for any person or persons to construct a bridge or to establish a ferry over Red River within three miles of the present location of Colbert's Ferry."[25]

The bill, which passed into law without discussion that fall, also guaranteed Frank Colbert exclusive franchise rights for the next quarter century.

With authorization in hand and protection in place, Colbert hired a Missouri engineering firm and set to work on his wagon bridge.[26]

Just as railroad engineers had found, technologies of the day would have a hard time spanning the Red River. The riverbed was sandy and soft with little shallow bedrock to support bridge piers, and the flat lands north and south of the river provided little elevation to raise the roadbed over the water. Worst was the ever-present danger of seasonal flooding that might bring tons of debris rushing downriver to beat against the bridge supports.

"Work on the bridge began in the spring but progressed very slowly," Colbert's employee recalled. "Travelers in wagons still kept coming, but we had no more freight wagons except a few."[27]

Colbert continued to provide limited ferry service during the year it took to complete his bridge; reduced ferry revenues and unexpected costs ate into his capital.

In August 1875 a local newspaperman inspected the completed bridge. "It is 577 feet long and 16 feet wide," he wrote, "with a turn-off in the center for passing teams 24 feet wide. It cost Mr. Colbert $40,000." The bridge was built on three piers of loose stone; the roadbed was of wood heavy enough to support the large freight wagons and teams that would cross it. A toll collector sat in a shed at midspan, collecting a dollar for every wagon drawn by four animals, seventy-five cents for wagons or carriages drawn by two animals, and twenty-five cents for a carriage with a single animal. A single horse and rider paid fifteen cents; a foot-passenger paid a nickel.[28]

Toll rates were steep, but a driver could walk his team across the bridge in a matter of minutes. There was no time lost waiting for the next ferry, no hassle trying to convince balky livestock to step onto a rocking ferry platform.

From the beginning, Colbert's bridge made money hand over fist. The railroad had siphoned some of the long-distance freight ferriage, but there was still plenty of local traffic for his new wagon bridge. A burgeoning new town, Denison, was growing up south of the river, and farmers in the Indian Territory used Colbert's new bridge to bring crops and livestock to market there. Other traders used the wagon bridge to cross north into the Territory.

Colbert's new bridge lasted just eleven months. A week of late summer thunderstorms in 1875 raked the Red River valley, sending walls of water and debris pouring downstream.

"The railroad bridge went out first," Colbert's toll collector said. "One span of it floated down and lodged against the north pillar of the wagon bridge, but did not even shake it. There was a heavy drift of logs and trees coming down and much of this lodged around the middle pier. Sometimes it would break loose, then big cottonwood trees would strike it endwise and bounce back like rubber balls."[29]

Continuous pounding eventually dislodged the rocks of the center pier. When it crumbled, the two center spans collapsed into the torrent, and Colbert's $40,000 investment washed away in a matter of hours.

"With the aid of a carpenter and some other help we had the ferry running again in about ten or twelve days," Colbert's employee said.[30]

Stung by the financial loss, Colbert returned to the ferry business while he called in engineers to build a replacement bridge. The new wagon bridge, completed in 1882, was again destroyed by flood three years later. Despite his wealth, Colbert couldn't afford to continue building (and losing) bridges. He chartered a corporation, Red River Bridge Company, and acquired some of Denison's most influential businessmen as investors.[31]

By 1891, bridge-building technology was beginning to catch up with the challenges of the Red River. Red River Bridge Company began construction of a new $50,000 iron-and-steel wagon bridge in 1890. The piers were anchored to bedrock and bumpered against pounding by debris; the spans were cast iron reinforced by steel. Builders also constructed a set of floating booms that would span the river at high water, shunting debris between the vulnerable piers. Flooding delayed construction, but the new(est) wagon bridge opened in September 1891.[32]

Within a month more than two hundred fifty paying customers were crossing the company's bridge every day: caravans of manufactured goods moving north into the increasingly populous Indian Territory; wagons full of household goods; carts packed with crates of fresh produce, milk, or eggs;

carriages of well-dressed families out for excursions; buggies and pushcarts; livestock and lone riders. By the end of the decade, the toll bridge crossing at Colbert's Ferry was the busiest crossing on the Red River, and everyone who traversed it paid a toll to Colbert's Red River Bridge Company.

By 1900 Denison had reached a population of more than eleven thousand, its growth fueled largely by its rail connections, a strong manufacturing base, and its just-across-the-river proximity to Indian Territory. Legislative wheels in Washington, DC, were already turning toward a merger of Indian Territory with the new Oklahoma Territory and the eventual creation of a new state north of the Red River. Denison boosters were ecstatic about their town's role as a gateway northward, but they expressed concern that bridge tolls might impede travel and trade.

Though Denison businessmen expressed their interest in "an organized effort to make the Red River wagon bridge free to public use," the bridge company wasn't about to forsake its cash cow. In fact, company investors saw a new opportunity on the horizon.[33]

In 1913, the officers of Red River Bridge Company announced plans to build a new bridge—a wider and stronger bridge—to replace the older one. The expansion, they said, was necessary to accommodate the new motor coaches and motor carriages. The Red River Bridge Company would build a new toll bridge for automobiles.[34]

CHAPTER 2

Mud to the Axles

RAILROAD millionaire E. H. R. Green is credited with driving the first production automobile into Dallas on Thursday, October 5, 1899. He arrived at 7:30 p.m. after a five-hour drive from Terrell, thirty miles away.

Describing the trip, he said that "cotton pickers dropped their sacks and ran wildly to the fences" to see his gasoline buggy fly along the dusty country roads at speeds approaching fifteen miles an hour. "At least a dozen horses executed fancy waltz steps on their hind legs as we sped by."[1]

Green, who operated the family-owned Texas-Midland Railroad from company headquarters in Terrell, had ordered the automobile from the St. Louis Motor Carriage Company. It had arrived by rail two days earlier, accompanied by a factory mechanic who would serve as driving instructor as he taught Green how to maintain the machine. A two-cylinder water-cooled engine was mounted under the wooden buggy seat, attached by a single chain drive to the rear axle, which turned two enormous bicycle-style wheels. Green started the car with a hand crank on one side and steered with a tiller that arched up out of the floorboard. The body panels were painted bright red with gold filigree trim.

"We didn't put on full power on the country roads because it would have been too dusty for comfort," Green said of his trip to Dallas. "When we struck the asphalt pavement on Main Street, we dared not do so because the thoroughfare was so crowded that it would have been dangerous to human life."

On his arrival in Dallas, Green spent several hours driving the gasoline buggy up and down lighted sections of Main Street and Ross Avenue, squeezing the horn bulb at every opportunity and enjoying himself immensely.[2]

At the beginning of the twentieth century, the automobile's greatest use was as a rich man's toy.

There is no authoritative date for the "invention" of the automobile—a self-propelled steam-powered military vehicle was demonstrated in France in 1768—but most date America's age of automobiles from 1900. Development of the internal combustion engine and storage battery technology in the late nineteenth century led an increasing number of carriage makers, bicycle repairmen, and mechanical tinkerers to try their hand at building a personal vehicle that carried its own means of propulsion.

A 1904 census of manufacturers counted 3,725 automobiles produced in the United States that year, comprising an equal number of steamers, electrics, and explosives. But with an average cost in excess of $1,200, the earliest production automobiles were out of reach for most shopkeepers, factory workers, and farm families. Nevertheless, automobile manufacturing plants popped up in almost any US city, town, or village where skilled craftsmen and a talented salesman could be brought together over a well-equipped workbench. Production levels rose, but car prices remained high to cover the costs of building these exquisite machines one by one. American factories manufactured 21,600 autos in 1905, and a Dallas newspaper reported that more than eighty machines were being driven on city streets there. Tulsa reportedly had more than two hundred automobiles (garaged, no doubt, behind many of the more fashionable Brady Heights homes).[3]

The early automakers vied with one another to engineer vehicles that were faster, more powerful, more dependable, more convenient, and more comfortable than those of their competitors. They sponsored entries in various national and regional distance races and endurance runs to prove the superiority of their product. But even as they provided increasingly luxurious vehicles to their carriage trade clientele, many auto manufacturers of that first decade recognized that there was a broader market for specialized vehicles: motorized jitneys could serve towns that couldn't afford to build a trolleybus system; shopkeepers might use a motortruck to deliver packages to customers' homes; dairy farmers could drive milk cans to the nearest rail depot while the raw milk was still warm; doctors could make more house calls and see patients more quickly.

Some automakers—notably Ransom Olds and Henry Ford—were exploring ways to manufacture these complex, bench-made machines without the use of skilled mechanics and expensive hand labor. Slashing labor costs and speeding up the manufacturing process would be pivotal in providing affordable automobiles for ordinary folks. Both men studied principles of mass production—interchangeable parts, continuous flow, division of labor, reduction of wasted effort—and applied them to the auto-making process. Olds's Curved Dash Oldsmobile and Ford's Model T demonstrated that reliable, quality autos could be produced at prices affordable to everyone.[4]

By 1913—when Henry Ford began manufacturing his Model T on a moveable assembly line and the Red River Bridge Company began construction on its new toll bridge—the automobile had ceased to be the sole provenance of the wealthy few and had become affordable to the masses.

Automobiles changed the nature of life in the twentieth century. Increasingly affordable gasoline autos became a favored means for transporting a family to church on Sunday, for driving across town to see friends, or for driving into the countryside for a summer picnic. Horses became scarce on city streets as businesses turned to the more efficient (and faster) motortrucks to deliver goods. On the farm, heavy-duty autos and trucks opened wider markets for produce, meat, milk, eggs, and fruit. Cars and motorized buses helped rural youngsters get to school while easing some of the isolation of farm life. Just ten years after that first motorized carriage chugged into Dallas, Texas had registered more than 14,000 personal automobiles; two years later there were 33,000. In 1911, Oklahomans boasted 2,000 autos in use and 12,000 by 1913. And the numbers continued to increase exponentially: in 1918 Texas had 194,000 vehicles registered; Oklahoma had 223,000 by 1922.[5]

Trouble was, there just weren't enough good roads for all those cars to drive on.

A decade into the age of automobiles, the *Daily Oklahoman* sponsored an endurance race, the first in Oklahoma to be sanctioned by the fledgling American Automobile Association. In spring 1910 the newspaper invited automobile manufacturing companies to pit their best new cars (and professional drivers) against one another in a contest of speed and durability over an eight-hundred-mile course. Factory-sponsored automobiles would run a loop route over the unimproved roads of southern and western Oklahoma, a route that would begin and end in Oklahoma City. Twenty auto manufacturers signed up to compete.[6]

Monday morning, May 29, 1910, at seven o'clock, sixteen showroom-fresh production automobiles lined up outside the *Daily Oklahoman* building at Fourth and Broadway. (Four cars broke down on the way to Oklahoma City and missed the start time.) The weather was perfect, skies blue and temperature moderate. A local debutante waved a flag to start the race, and the cars roared off toward the Western Avenue bridge and south toward Chickasha.

By end of the first day two more cars were out of the race: one with a sheared hub from hitting a stump hidden in the roadbed and the other with an engine fouled and ruined by dust. The next morning two racers became mired in the quicksand of a creek bottom and required teams of mules to haul them out.

The next five days were a litany of blowouts, smashups, broken axles, and fouled engines as the competitors raced south to Fort Sill, westward to Hobart, then northward toward Alva, Enid, and Ponca City. (The racecourse was adjusted on the fly when judges deemed the gyp hills of northern Oklahoma too steep and too corrosive for the autos that remained.)

The first of six finishers completed the eight-hundred-mile route and crossed the start/finish line after sundown six days after starting out. Three of the six finishers were disqualified after judges inspected driving logs and determined that the autos had strayed more than forty miles off course because of the lack of road markings.[7]

These distance and endurance races captured the public's imagination while showcasing vehicles from competing manufacturers. Nevertheless, seventeen of twenty of the finest production automobiles in the United States entered in this race were beaten by Oklahoma's rough, rutted, dusty, potholed, unmarked rural roads. And it wasn't just Oklahoma; most rural roads remained more suited to a slower, horse-drawn way of life.

A federal survey of the United States in 1904 mapped just 141 miles of paved roads outside cities. Turn-of-the-century rural roads—especially in Texas and soon-to-be Oklahoma—were most often crude earthen pathways, zigzagging from community to community, circling around any grade steep enough to slow a two-mule team, and meandering along creek banks toward the likeliest low-water crossing or ferriage. Bridges, if they existed, were homemade affairs, able to bear a loaded farm wagon but likely to collapse under the weight of a two-thousand-pound automobile. Winter snow and ice might bury a rural road for weeks at a time; spring rains and runoff could gouge out large sections. (The farmer in a carriage or wagon might find a path around these washouts; the motorist would find himself bogged down in mud up to the axles.)

Rural roads before 1900 were strictly local affairs, maintained by the farmers whose property they passed or under the oversight of county authorities. Standardized directional signage? There was none; the locals didn't need it. As long as natural horsepower could get the farmer's crops to the nearest town or rail depot and his family to church on Sunday, there was little demand for expensive road improvements.[8]

As the nineteenth century turned into the twentieth, however, three forces—cyclists, farmers, and auto owners—combined to spark public demand for road and bridge improvements, a century-long program that stands as one of the largest improvement projects ever undertaken in the United States.

Oddly enough, bicyclists were among the first to begin agitating for better rural roads. The League of American Wheelmen was a national organization of recreational cyclists and sportsmen advocating the natural and healthful benefits of bicycle touring, and bicycle touring depended on a network of well-maintained country roads. The American Wheelmen's efforts resulted in a smallish congressional appropriation for the creation of an Office of Road Inquiry (ORI) within the Department of Agriculture. Constitutional concerns and a general resistance by states' rights congressmen (many of whom were rural southerners) limited the ORI's activities to data collection, education, and promotion.[9]

A second force was the demand for rural mail delivery. Farmers and ranchers of Texas and the Twin Territories (and the rest of the rural United States) had no choice but to ride (or walk) over rude trails each week or so into the nearest town with a post office. There, they would collect their letters, bills, magazines, catalogs, patent medicines, and everything else that city folks could expect to have delivered to their doorsteps. After heavy lobbying by the Farmers' Alliance (and mail-order retailers such as Sears, Roebuck and Co. and Montgomery Ward), Congress funded a small rural mail delivery test in 1896, and by 1905 a small army of 32,000 Rural Free Delivery carriers were riding back roads daily. The poor quality of those roads led rural carriers to threaten their patrons with a mail stoppage if they didn't maintain the roads to their post boxes. The Post Office became a vocal lobbying force for route inspection and the improvement of rural roads and bridges.[10]

Wheelmen and postmasters sang from the hymnal, but autoists were the most successful at proselytizing on behalf of the good roads in the United States. Automobile ownership ignited a national movement advocating better roads.

The National Good Roads Association, founded in 1900, was an offshoot of the cyclists' organization, but was quickly embraced by the burgeoning population of auto enthusiasts. Autoists were joined by the voices of farm groups, railroads, businessmen's organizations, and civic groups in propounding the good-roads gospel, stressing the economic and social benefits of better roads. The association became a crusade as it spawned more than a thousand state and local chapters from coast to coast; these regional chapters sponsored conventions and public demonstrations, and often became involved in local politics to promote road improvement.[11]

A particularly destructive series of floods across Oklahoma Territory in 1902 spurred A. C. Titus, a fruit grower from Crescent, to organize the Sand Valley Good Roads Association, the first in the Territory. Titus encouraged the formation of other groups and in 1904 chaired the first convention of the Oklahoma-Indian Territory Good Roads Association in Guthrie, the territorial capital.[12]

When the Dallas Commercial Club (a sort of cross between a chamber of commerce and the Rotary) hosted a national Good Roads officer at their weekly meeting in 1904, the event became more tent revival than business luncheon. Roused to action by the luncheon speech, the members spontaneously voted to organize and sponsor a statewide Good Roads convention within ninety days.[13]

Business leaders of Denison held a regional Good Roads Association organizing meeting in February 1905. (Attendance was disappointingly low, however, as bad weather and bad roads kept many out-of-towners from attending the discussion of better roads for the area.) Attendees passed a set of resolutions expressing their judgment that "the present method of road work and road construction in this county is barren of results in the way of permanent roads." This, despite "annually using large amounts of the people's money." The organization was unanimous in urging that "some system of road construction be adopted that will give us permanent and lasting highways."

While citing the toll bridge at Colbert's Ferry as an impediment to free travel, attendees also recognized the importance of easy access to the Red River bridge: "We favor an organized effort to make the Red River wagon bridge free to public use, and we urge the improvement of all roads leading thereto on both sides of the bridge."[14]

Abilene to Ardmore, Tulsa to Tyler—towns and communities throughout Texas and the Twin Territories hopped on the Good Roads bandwagon, adopted the "Organization, Agitation, Education, and Legislation" motto, and pledged themselves to transform the packed-dirt rural wagon trails of their area into broad, firm, all-weather roads that would help farmers transport goods to market faster and at better cost. The local Good Roads Association groups provided a national platform to help increase public awareness, promote citizen involvement, and lobby state and local governments to help fund road improvements.

Better roads might be a laudable aspiration, but who would pay for them?

Texas and the new state of Oklahoma had plenty of roads in 1909: according to an ORI survey, there were 71,300 miles of road in Oklahoma and 128,900 miles in Texas. Calling them "roads," however, might have been a little generous. Combined, the states had only 6,600 miles of improved roadway, most of it in or near the large cities. ("Improved" might mean as much as a brick-paved Main Street or as little as a mule-graded country road with a load of light gravel crushed into the earthen surface.) That left 193,600 miles of ruts across the alkaline desert, packed mud tracks through the forested country, sandy trails through farmlands, and miles of open prairie land pounded hard and flat by decades of hooves. While other states spent $1,500 or more per mile to grade and gravel their unimproved roads, Oklahoma and Texas were spending next to nothing.[15]

Texas, particularly, was slow to invest state money in road improvements; roads were considered local matters, the responsibility of those who needed to get from one place to another. Legislation in 1879 allowed county governments to "draft" the services of all males aged eighteen to forty-five for annual service of up to ten days to build, repair, and maintain county roadways. Extremely unpopular, the "road labor tax" generally worked well in the populous counties but left the sparsely settled areas able to do little more than hitch a few mules to a split-log road drag and pull it over the busiest roads after spring rains. Constitutional amendments in the 1890s allowed counties to collect road taxes and use convict labor for roadwork. County control of road building was entrenched further in 1903 with another amendment that gave county courts the power to issue bonds for local road improvements. County control made for inconsistent construction standards and a road grid that didn't necessarily serve the best interests of intrastate and interstate road travelers. It would take the state of Texas almost a decade and a half—the critical first fifteen years of the auto age—to wrest control of highway building, improvement, and maintenance away from its 254 county governments.[16]

The state of Oklahoma, on the other hand, was born into the Good Roads movement. Successful lobbying during Oklahoma's 1907 constitutional convention resulted in the establishment of a state highway department. The department wasn't established by the legislature until 1911, when the governor appointed Sidney Suggs, a noted Good Roads booster, as its first commissioner. Suggs's fledgling department was given no appropriation for construction or maintenance budget—county commissioners remained in charge of those purse strings—but in his first years Suggs and his four employees sketched out a state highway system, tested methods of road construction and shared results with the counties, coordinated the efforts of local Good Roads associations, and educated legislators about the financial benefits improved roads could bring to their constituents.[17]

Left to the individual counties, the costs of road and bridge improvements would be staggering and the progress glacial. Meanwhile, motorists were demanding better highways, and farmers wanted more dependable, efficient routes to their markets. A new automobile toll bridge at Frank Colbert's old Red River ferry crossing would provide both.

A 1908 flood once again washed away the Red River toll bridge, but the Red River Bridge Company investors were in no hurry to replace it. A new bridge—especially one suited for automobile traffic—would require (another) sizable investment. Also, some local farmers had jerry-rigged a wagon shelf off the side of the Carpenters Bluff rail bridge, eight miles downriver, siphoning off ferry revenues. It was apparent to the Red River Bridge Company directors, however, that the automobiles were here to stay. Every day brought more motorized vehicles to the ferry crossing: family vehicles, delivery vehicles, farm vehicles, excursion vehicles. Some were too heavy (or too large) to fit the ferries, and there were reports of heavily loaded autos crashing through the framework of the wagon bridge at Carpenters Bluff.

Most of the bridge company directors were Denison and Sherman businessmen, and they recognized the increasing need for unimpeded travel across the Red River. Local boosters saw Denison as the "Gate City," a commercial stopover for travelers on a proposed new Canada-to-the-Gulf transportation route. Even the railroads were lobbying the bridge company to build a new automobile bridge, knowing it would open the way for more freight to arrive at (and be shipped from) the Denison freight depot.[18]

A railroad man of another sort, J. F. Strickland of the Dallas-based Texas Traction Company, approached the bridge company with an offer. Texas Traction operated seventy-seven miles of electric interurban train lines that connected Dallas, McKinney, Sherman, and Denison, and the company desired to expand northward. Strickland asked the Red River Bridge Company to build a bridge and engineer it to accommodate electric rail lines. In return, Strickland would buy fifty-eight acres of Colbert land owned by the bridge company for future use as an electric powerhouse. On January 25, 1913, directors of the Red River Bridge Company approved a $50,000 contract to build a new bridge across the river north of Denison.[19]

The *Dallas News* called it "magnificent" and "perfectly suited in every detail." By the time it opened for traffic on March 1, 1915, the new Red River toll bridge cost somewhere north of $100,000, but engineers deemed it the most solid, state-of-the-art bridge anywhere along the river. The 675-foot-long bridge consisted of two major spans, supported mid-river by a steel pier wrapped six feet thick with more brick and concrete. The pier stood on a footer that, itself, rested on bedrock nearly fifty feet below the river surface, and curved steel bumpers diverted heavy debris that might rush down the river at high water. The two main spans of the superstructure were framed by trusses with a concrete deck that had been cast in place. The bridge was two lanes wide and capable of carrying an eight-ton interurban rail engine and cars. The roadway was eight feet higher than the highest recorded river level; twenty-foot-high railings lined the approach spans, which connected abutments to the main spans. Collectors in neat tollbooths on both ends of the bridge could communicate with one another using a signal block system.[20]

Magnificent as the bridge may have been, some still groused that it was a private venture, collecting twenty-five cents from every individual who drove, rode, or walked across the bridge. *Southern Good Roads* scolded Texas, Oklahoma, and other states for their lack of road improvement progress, warning that the revival of turnpike and toll bridge companies "can only be approved as a protest against local slowness." As good as they are, these pay-for-travel bridges should disappear as soon as possible, the magazine argued, "for all the roads should belong to all the people and be paid for in equal degree by all the people."[21]

True enough, shareholders of the Red River Bridge Company put up their own money to build the bridge, and those same investors stood to make a good profit on their investment over the years. In 1915, neither Texas nor Oklahoma was finding it necessary (or practicable) to make the same investment. Adjoining county governments could have borrowed money to buy out the bridge company's interests, but instead allowed private investors to foot the construction and operations bills.

Besides, the Red River Bridge Company paid to build more than its own bridge; it spent substantial sums on the approach roads. The company paid to macadamize the county road from Denison, to pour concrete culverts, and to landscape with Bermuda grass. For the north end of the new bridge, the company set aside $5,000 to build a new road through the town of Colbert toward Durant. It paid to cut a gently inclined path up through the sand hills on the north riverbank (using the sand to mix concrete for the bridge), tamp the sand pathway down to a hard-packed road surface, roll limestone chips into the roadbed, and then drench the road surface with asphaltic oil.[22]

Good roads and a good bridge meant good business for both the toll bridge investors and the community as a whole.

Sidney Suggs, state highway commissioner of Oklahoma, assured Good Roads advocates in Denison that the new bridge "would attract thousands of visitors and tourists this way." Denison merchants had already recognized the possibilities. They were positively gleeful about the new bridge and easier access into southern Oklahoma, boasting that they would be taking retail trade from Durant. Saturday had been the traditional Trade Day in Denison; after the new bridge opened, traffic jams developed downtown, forcing retailers to designate Wednesday as a secondary Trade Day for all the out-of-towners who came to buy and sell.[23]

The new automobile toll bridge opened markets for businesses south of the river, but it also provided market advantages for farmers north of the river. Prices for feed and supplies dropped throughout the region as merchants in Denison, Durant, and Colbert competed for the business. An editorial in the *Daily Oklahoman* described the financial advantages to farmers who—with access to good roads and sound bridges—could bring their products to market when demand and prices were highest, not needing to wait for gins in Colbert, Denison, or Durant, or whichever gin had time available and was offering the best price.[24]

For toll bridge investors and other Denison business interests, the greatest benefit of the new bridge was that it put the Denison-to-Durant Red River crossing into the running for one of the developing transcontinental highway routes. As more US families ventured out on motoring vacations, civic boosters would lay out a route, give it a marketable name, collect dues from businesses and towns along the way, and then promote the route to motorists with prominent road signage, trail logs, maps, and discount booklets. The Meridian Road, proposed by a Kansas promoter as early as 1911, would run from Winnipeg, Canada, to Galveston on the Gulf of Mexico. The two-thousand-mile route would cut through nine states, roughly bisecting the United States and costing some $6 million (at the current cost of $3,000 per mile). Grayson County (including the towns of Denison and Sherman) announced plans in 1912 "to try to induce the promoters to build through here."[25]

Denison and Sherman were too late. The Denison auto bridge was not yet complete and Texas and Oklahoma counties were still struggling to improve the main roads. The Meridian Road promoters found a better route to the west.

In 1917, however, with the bridge complete, Denison joined with Tulsa, Dallas, McAlester, Houston, and associations in seven other states to promote the new King of Trails Highway. It was a good bet. The King of Trails Highway—extending from Winnipeg to Galveston—became one of the most popular of the named interstate highways of the 1920s, spawning a healthy roadside infrastructure of campgrounds, diners, motor hotels, gasoline filling stations, tourist attractions, auto repair shops, and more.[26]

By 1930 more than 500,000 travelers a year would follow the King of Trails Highway to the Red River, each automobile, truck, motorcycle, and wagon paying a toll to cross the Red River Bridge Company's bridge at Colbert's ferry.[27]

CHAPTER 3

An Honest Oilman

UNTIL PASSAGE OF the Federal Aid Road Act in 1916, road improvement progress in Texas and Oklahoma was excruciatingly slow and predictably wasteful.

County governments held the purse strings, and a desire for better local roads ran head-on into the reality of paying for them. Some counties couldn't afford (or chose not to spend money on) rural road improvements, leaving gaps in the statewide systems. Road-building and maintenance jobs were considered mere rake-and-shovel work, so counties rarely hired skilled workers or road engineers, leading to roads that lasted barely a rainy season before washing away. Some counties purchased a full complement of road-building equipment, whether or not the equipment would be fully utilized throughout the year. There was petty graft, too. The owner of the local gravel quarry or the person hired to maintain county construction equipment was often an in-law of the county judge or sheriff. Road jobs provided local officials a juicy opportunity for patronage. Texas and Oklahoma counties were allowed by law to use convict labor on road crews, leading to further abuses of power.[1]

Oklahoma, at least, had a working state highway department to coordinate and consult with officials of its seventy-seven counties. In 1915, the state legislature provided for the hiring of a state engineer and an assistant. Further, the state added a road tax and a charge for auto registration, the monies to be funneled to the counties for road improvements. In 1917 the Oklahoma legislature voted a million dollars to be expended on state roads by county officials in accordance with plans provided by the state highway engineers.[2]

In 1917, Texas became the last state in the nation to establish a state highway department, and did so only after being coerced by the federal government. The Federal Aid Road Act promised road-building funds to states, but only if the state had an organized highway department to provide supervision and competent engineering assistance. Most Texans were reluctant to surrender responsibility for road construction and maintenance to a faraway state capital, and the state legislature provided the new highway department with the bare minimum of funding. However, the Federal Aid Road Act was amended in 1921 to offer matching federal funds to supplement state money—but only if the state highway department had *exclusive* control over road building and maintenance.[3]

Like it or not, the highway departments in Austin and Oklahoma City would be calling the shots and letting the contracts on state highway construction. Construction contractors, road equipment manufacturers, and lobbyists began packing their bags and hopping trains for the state capitals.

In 1924, the Oklahoma Highway Department put $9 million a year in road and bridge contracts out for bids; Texas awarded $25 million, an amount greater than the state spent on higher education, state hospitals and care facilities, and the state judiciary combined.

And where money flows, temptation grows.

Rumors of corruption—vanishing equipment, sweetheart contracts, overpricing, and kickbacks—in the Texas Highway Department swirled throughout the contracting industry, but it wasn't until August 1925 that Louis W. Kemp, secretary of the Texas Highway and Municipal Contractors Association, could build a case. Based on a tip, Kemp found that a small come-from-nowhere general road contractor from Houston was using a state tractor and roller on a private job. Sherman-Youmas had won some minor city and county grading jobs, but then had been awarded a large state contract for maintenance and improvement of the highway between Houston and Galveston. The contractor was switching out state equipment on private jobs. A review of the state contract showed some curious provisions: dates, materials to be used, pricing, and (most glaring) the absence of competitive bidding.[4]

When Kemp attempted to discuss the irregularities with state highway officials, he was rebuffed; two state highway commissioners refused to speak to him at all. Kemp took his facts to Texas attorney general Dan Moody. It didn't take the state attorney general long to discover a $3,500 check written by the owner of Sherman-Youmas to a highway commissioner. Moody dug deeper.[5]

In 1925, Dan Moody was a thirty-two-year-old red-haired straight arrow who had built a reputation prosecuting Ku Klux Klansmen. He was elected state attorney general in an administration many considered to be the most corrupt in Texas history. Moody's investigation turned up a dozen cases of highway contracts let to friends of two highway department commissioners without competitive bidding and at exorbitantly high prices. Moody launched a crusade against the corruption, suspending contracts, impounding funds, collecting settlements, and hauling the biggest offenders into court.[6]

"They bored a hole in the treasury of Texas," he told one jury, "and before the leak was stopped nearly two million dollars had fallen into the hands of the American Road Company and more than a million dollars had been paid to the Hoffman Construction Company."[7]

Angry newspaper editorials called for the firing of the state officials responsible: "A new highway commission is needed immediately and the men appointed should be of outstanding ability." Two of the commissioners resigned in disgrace, but the governor refused to condemn the mess.[8]

Sick of crooked deals, corruption, and political cronyism, Texas voters threw out the old administration and elected Moody governor in 1926. As one of his first appointments, the new governor chose Ross S. Sterling, a squeaky-clean Houston businessman, to chair the Texas Highway Commission.

Although his adult careers would take him into boardrooms, ballrooms, and the governor's mansion, Ross S. Sterling was born a country boy.[9]

Sterling was the eighth of twelve children, born in February 1875 on a farm in the Double Bayou community near the Gulf Coast town of Anahuac, southeast of Houston. His father was a Confederate veteran; his mother was the daughter of his father's commanding officer. Family names having run out by the time he came along, Sterling was named after Confederate general and legendary Indian fighter Sul Ross. He would be an active supporter of Confederate veteran causes for the rest of his life.

Like his father and brothers, he worked the farm, feeding livestock and tending crops. Leaving the farming to his boys, the elder Sterling opened a small dry goods store. Between crops, Ross was learning his letters at a one-room schoolhouse, and proving to be not much of a student. When Sterling was thirteen years old his mother died, and he never returned to school.

It was while working on the farm that Sterling learned an important life lesson. "I discovered that by exerting a little extra effort and enterprise I could get ahead," he said. "If I hit four licks while the other fellows were hitting two or three, they just can't keep up with me. That became my life's philosophy."[10]

The Double Bayou community was isolated. Market roads, if they could be called that, zigzagged along levee tops or bogged down completely in the marshy country. In 1895, Sterling and his brothers built a forty-six-foot cargo sloop and began freighting farm goods and livestock to markets across the bay in Galveston. The Sterlings brokered local vegetables, corn, cotton, melons, chickens, eggs—even horses and cattle—to Galveston buyers. The family brought back clothing, furniture, tack, and other manufactured goods that they sold in their Double Bayou store.

Eventually, Ross's father put him to work in the family store. He found the daily routine of storekeeping to be tedious and confining—especially for an active outdoor youth—but he also discovered he had a special knack for trade and deal making. Saving the ten-dollar-a-month salary his father paid him, Sterling was soon able to afford to marry and to acquire a small store of his own farther down the bayou.

In 1901, Sterling sold his store for $4,500. "I was a twenty-six-year-old country boy," Sterling said. "That was a small fortune in my eyes."[11]

Sterling had heard of the business frenzy occurring in oil boomtowns of southeast Texas. In 1901, the Spindletop well came in, the first well of what would become the world's highest-producing oil field to date. A gold rush mentality enveloped the small towns surrounding Beaumont, where swarms of speculators were looking for leases, and crews of roughnecks were looking for equipment and supplies.[12]

Drillers required plenty of dray horses and mules to haul bits, line, and framing timber, and the horses and mules needed to eat. The country boy from Double Bayou went to Sour Lake, a little village in the piney woods, and used some of his $4,500 to open a feed store. Sterling had grown into a big man—six feet tall and more than two hundred pounds—and he could carry a hundred-pound sack of oats under each arm from his wagon to his buyer's barn. Sterling's wife operated the feed store while he was up at dawn making deliveries. Hitting four licks where his competitors were hitting only two or three, he soon expanded to own four stores, offering credit and managing his own accounts.

Cash from the feed stores allowed Sterling to invest in other businesses near the oilfields popping up east and northeast of Houston. He opened several sawmills and, with four partners, established a fifteen-thousand-acre rice farm (on land where oil would be discovered several years later). The Panic of 1907 slowed the pace of business for a while, but allowed Sterling to take controlling interest in four small-town banks. He bought land in Houston and began developing a luxury subdivision.

Sterling shrewdly leveraged his money. He bought or built businesses on borrowed money, and when he sold he generally retained a piece, a small share that would continue to pay dividends year after year. At thirty-six years old, the country boy was approaching rich, but he wasn't yet oil rich.

Ross Sterling had followed the boomtown parade for a decade, but he had never invested in the oil business. Doing so would be too risky for a family man, he believed. Wildcatters might spend every penny they could scrape together to lease land, hire a crew, and build a rig, only to drill a dry hole. Even a proven, productive well pumping a hundred barrels a day might play out tomorrow. (It was a time "when witches' wands were more reliable than petroleum geologists," he recalled later.) Still, Sterling saw men of lesser abilities become overnight millionaires. In 1910, he finally took the plunge.[13]

The Humble oil field had come in big, but nobody knew how long it would last. A pair of wildcatters owned two wells, each producing fifty barrels a day, but they wanted to cash out and move on to the next boomtown. The price was a modest $12,500. If the wells continued to produce, they would pay out in less than six months. Sterling bought the wells, giving half interest (and operating responsibility) to an experienced oilman he knew from Sour Lake.

"Having got my feet wet in the golden grease, I soon waded out deeper," Sterling told a friend.[14]

Sterling's trading skills, access to cash, willingness to assume risk, and straightforward honesty in the conduct of business made him a standout among the boomers, adventurers, gamblers, confidence men, and other shady characters who populated these early Texas boomtowns. He continued to acquire producing wells in the Humble field at a bargain, also collecting leases, land, and equipment as he went.

Within a year, he said, "I felt that I could call myself an oilman."[15]

In 1911 Sterling gathered several other local oil producers and proposed they form a public oil production company. With a combined $150,000 investment, the group was awarded a charter as the Humble Oil Company, with Ross Sterling as president.

The growth of the company was breathtaking. Sterling reduced the dividend to 1 percent, and used the surplus to buy additional production. Leveraging the surplus, he exchanged stock to acquire ownership of exploration companies and drilling contractors in Texas and Oklahoma. Humble Oil Company drilled its share of dry holes, but it hit enough two-thousand-barrels-a-day gushers to more than make up. As the amount of crude that Humble Oil produced grew, Sterling looked for ways to increase the amount he was paid for each barrel. He was currently selling to the larger Texas oil companies who were, in turn, reselling at a higher price to the East Coast refineries. Sterling organized the smaller Texas producers into the Gulf Coast Independent Producers Association, giving them all enough combined volume to sell directly to the easterners for the higher price. In 1917, Sterling gathered all the associated Texas producers and proposed that they reorganize into a single large company. The larger company, Sterling said, could afford to build its own refinery and pipeline capabilities. Once again, his salesmanship and ability to make deals won the day. Sterling went to Austin to obtain a new charter for the reorganized Humble Oil & Refining Company. Ross Sterling, of course, was named president.[16]

Sterling oversaw the purchase of land in Goose Creek, Texas, on which the company would build the largest oil refinery ever constructed. He leased lands for pipelines to run to the new oil fields in West Texas, Wichita Falls, and Ardmore. But Humble Oil needed cash for all this expansion (and it was in Sterling's nature to pay for his risk with borrowed money).

Standard Oil of New Jersey, one of the old Rockefeller companies and the largest oil company in the United States, contacted Sterling with an offer to purchase a majority of Humble and provide financing for expansion. Sterling and his directors traveled east to discuss the deal.

"We have many stockholders in Texas," Sterling told the Standard Oil president, "and a lot of them don't think as much of you fellas as we do." He turned down the deal and the $17 million Standard was offering. In the end, Humble agreed to sell 50 percent—exactly half—of the company for about $15 million. Knowing that Standard was uncomfortable without full control, Sterling assured the company chairman: "Any time you think we're not doing things right down there, I'll transfer you some of my stock."[17]

They took Ross Sterling at his word.

Using Standard Oil money, Humble built its refineries and pipelines. In 1919, Humble opened its first retail gasoline filling station in Houston, even as it began setting up branded gas and oil distributors throughout the Southwest.

Ross S. Sterling, the unlettered country boy from Double Bayou who started out as a $10-a-month grocery clerk, controlled the second-largest oil company in the United States. He was more than rich; he was oil rich.

In the years since Double Bayou, Sterling had acquired more than money; he had acquired a reputation as an honest businessman. He was not well educated, but he was shrewd; his handshake was his word and his deals got done. Instead of traditional schooling, Sterling's education came from the primer of his business experience. He learned that private enterprise can whip almost any problem but government. Play the long game, be patient, and know that the men of government will eventually see reason. He also developed the entrepreneur's conviction that any person can get ahead—any person can be Ross Sterling—if only he has the fortitude, the drive, and the willingness to work hard.

In 1924, Sterling hosted a dinner for his Humble Oil directors and senior executives.

At slightly more than six feet, he was taller than most of the men enjoying cocktails in the bar before dinner. He retained the burliness of his youth, and then some, although he carried his 275 pounds high up on his frame. Owing to time spent outdoors, his complexion was ruddy, the color of a Jonathan apple. His sandy hair had turned to iron gray, and his blue-gray eyes looked out through a pair of plain spectacles. Ross S. Sterling had the appearance of a healthy man full of life and vigor. That's why his announcement to the men gathered for dinner was so surprising.[18]

"I'll be fifty in February," he told them, "and I'm pulling up stakes. I've worked as much in my lifetime as most men do in sixty-five years. Why keep my nose to the grindstone?"[19]

True to his word, Sterling resigned from Humble Oil in February 1925 and sold a large block of his shares. But no one who knew him expected Ross Sterling to sit in a rocker for the rest of his days. He still owned pieces or shares of almost every company he had touched—sawmills, stores, housing developments, flour mills, banks, a short line railroad, the first automobile parking garage in Houston—and he remained involved in those businesses. At the same time he bought, improved, and donated to the YMCA a site for a boys' camp, provided financial support to Texas Christian University, and chaired the Houston-Harris County Channel Navigation Board.[20]

Even as he was laying aside his Humble Oil responsibilities, however, Sterling became involved in the newspaper business. *The Dispatch*, a morning newspaper competing with the well-established *Houston Post*, was having a hard time making a go of it. The owners approached Sterling for help making payroll in return for their receivables. In no time, he had poured in $20,000; six months later he took over ownership of the troubled business. Typically, Sterling doubled down on his investment, raiding the staffs of the *Post* and the *Chronicle* while building a new production facility in downtown Houston. Soon it was the *Post* that was hurting as *Dispatch* circulation and advertising increased weekly. Never shy about acquiring a weakened competitor, Sterling bought the *Houston Post*, merged it with the *Dispatch* to create the new *Houston Post-Dispatch*, and hired former governor William P. Hobby to run the operation.

The *Post-Dispatch* developed a strong editorial voice; Sterling and Hobby were vehemently anti-Klan and disgusted by the corruption apparent in the office of Texas' governor. When crusading Texas attorney general Dan Moody herded the crooked highway contractors into a courtroom and forced the resignation of two highway commissioners, the *Post-Dispatch* trumpeted his efforts. And, in 1928, when Dan Moody ran for governor, Sterling put his purse and paper behind him.

"As a reward for my assistance," Sterling said, "Moody appointed me to the most important and the hardest position he had to offer: the chairmanship of the State Highway Commission."[21]

Having won the 1926 Texas gubernatorial election by rooting out corruption in the Texas Highway Commission, Dan Moody needed someone to rebuild what he had torn into and to do it honestly. He got plenty of advice from newspaper editorialists throughout the state: "Our idea is that the Highway Commission should be composed of the best business men possible," the publisher of one major newspaper wrote. "They should be allowed absolute freedom in the formulation of their policies."[22]

Ross Sterling was a savvy political choice as chairman of the Texas Highway Commission.

First, Sterling was a businessman. The Texas Highway Department was a $25 million business and growing every year. It employed 1,700 people in varied departments working on projects scattered throughout the state.

Sterling had the experience to manage that kind of business, and he was available to work for government wages. Second, Sterling had a statewide reputation. His wasn't a name often brought up at the breakfast table, but Sterling's involvement with TCU had him rubbing shoulders with a number of wealthy North Texans. He was known in West Texas for his oil and cattle interests, and his newspaper ownership in Houston gave him a big voice in southeast Texas. Sterling was known for his honesty. He was an aggressive dealmaker, yes, but with no tolerance for underhanded business.

Perhaps most important to governor-elect Moody was that Ross Sterling was an honest businessman from the *oil* business. Texas had recently enacted a one-cent-per-gallon gasoline tax to help pay for road improvements and maintenance. It wasn't enough, and Moody wanted to increase the tax by two, three, or four cents. Texas oil refiners had lobbied against the gas tax, but Moody felt Sterling could help oil producers and marketers see the benefits of increased road-building budgets.

Typically hitting an extra lick, Sterling went to Austin to look through the department before the other two commissioners arrived. "The highway department was in even worse shape than I had expected after the scandals," he observed. Construction was at a standstill; maintenance of state roads was virtually nonexistent. Much of the highway department's equipment was worn-out junk, left to rust where it had last been used. Records were chaotic (if they existed at all), and he saw distinct signs of malfeasance. Worst of all, Texas had been cut off from federal aid, a loss of $6 to $8 million a year.[23]

Sterling first called the Harris County auditor, persuading him to leave Houston and come to Austin. Together, Sterling and the auditor straightened out the books and installed a new bookkeeping system. Carrying a complete statement of the department's finances, equipment, and projects, Sterling went to Washington, DC, where he presented it to the chief of the Bureau of Good Roads. Sterling promised to provide regular audited statements of condition to the bureau, and federal money was soon flowing again.

Sterling held private meetings with heads of the major oil companies, showing in dollars and cents how an increase in the gas tax would help pay to complete more roads, which would increase miles driven, which, in turn, would increase the sales of gasoline products at retail. The oil companies agreed to help Sterling lobby the state legislature. In April 1927, less than four months after the new administration took office, the legislature voted an increase in the gas tax to three cents per gallon, largely the result of Sterling's efforts.[24]

New gas tax revenues allowed Sterling to replace the department's rusted-out equipment. With Sterling as chairman, the three commissioners held open hearings for every piece of equipment purchased and every contract awarded. When not in meetings, Sterling was driven in the department's big Lincoln over Texas back roads and highways, sometimes up to five hundred miles in a day. He would often stop to meet with roads crews to inspect their work and build morale.

In less than a year, dirt was flying. All over the state, roads were graded, culverts were built, and long stretches of highway were paved or topped. Sterling arranged for every newspaper in the state to receive a weekly Roads Report bulletin, with county-by-county descriptions of improvements, construction, closures, and completed roadway.[25]

Ross Sterling's name was becoming known throughout the state; his rejuvenated highway department was getting Texas motorists out of the mud and onto the road. But there was a brewing problem that the Texas Highway Department (and the Oklahoma Highway Department, for that matter) would find more difficult to resolve: toll bridges.

Oklahoma and Texas were making great strides in building out their state highway systems, but as the number of miles of all-weather roadway increased, so did public resentment of toll bridges.

"What's the idea of building a hard-surfaced highway with state money to the very entrance to the [toll] bridge?" an Oklahoma motorist asked. "Are we trying to drum up business for the operators of the bridge?"[26]

How can it be, motorists were asking, that we pay taxes to assure fast and efficient road transportation and still have to pay tolls to private individuals?

One Texas newspaper editor observed: "The motorist is disagreeably surprised, to put it mildly, that [the state] has not purchased these bridges and opened them up as part of its regular road system. They give the impression the state cannot afford to build its own bridges, or that public officials are in harmony with individuals who promote the bridges for their own profit."[27]

The expense and annoyance of a toll bridge was nothing new to those who depended on tolled crossings for their livelihoods. The rebuilt Red River bridge—so welcome when opened in 1915—was, just a decade later, considered to be an impediment to Denison merchants: "It is asserted that the present toll system is a pronounced drawback to Denison in that prospective buyers object to paying the toll in order to come here to make their purchases," the mayor said.[28]

For special events—county fair days or the Christmas shopping season—merchants might negotiate with a bridge owner to forgo toll charges. The merchants of Lexington and Purcell, Oklahoma, purchased books of a hundred passes and handed them to customers to lessen the inconvenience of the toll charges across the river between their towns. Eufala merchants announced proudly in their ads that "through our efforts the Jefferson Highway toll bridge has been opened for free travel."[29]

Local businesspeople feared that the toll bridge could impede Denison's reputation as Texas' Gate City, depressing tourist and trade revenues that flowed through the city. Tourist revenue was becoming increasingly important to Denison merchants. A second named (and heavily promoted) interstate route, the Jefferson Highway, ran from Winnipeg through Durant and Denison on its way to New Orleans. A report to the city's commercial club claimed that "fifty-seven percent of the 67,000 tourists entering Texas from other states during the last season entered by way of the Jefferson Highway through Denison."[30]

There was scarcely a resident within ten miles of the river crossing who wouldn't be better off with a toll-free bridge over the river.

"When the good roads movement gained considerable proportions it was assumed that the toll bridge, like the old-fashioned toll road itself, would disappear," according to an editorial in *The Oklahoman.* Far from disappearing, the number of toll bridges on state and federal roads was actually increasing.[31]

A 1927 survey by the American Automobile Association found 425 toll bridges in operation or under construction across the nation. Of that number, 217 were on roads that received federal aid and 60 were on the state highway systems. The growth was occurring because "there was easy money in toll bridges," according to Thomas MacDonald, chief of the US Bureau of Public Roads. Congress had been generous in granting charters to private companies in the belief they aided improvement of highway systems. The generosity spawned a number of "franchise grabbers" who staked out strategic positions where trunk highways were planned, built toll bridges, and began collecting the profits.[32]

And profits could be huge. The American Automobile Association cited several bridges that returned more than three times their cost each year.

"It is nothing unusual to find toll bridges earning all the way from ten to one hundred fifty percent." The organization—which had become a strong national lobbying force for motorists—vowed to ask Congress to "stop this extortion by the toll bridge trust."[33]

Toll bridge franchises were awarded locally, too. Uncertain as to when the Texas Highway Department might spend money for a bridge, Anderson and Navarro county commissioners invited private capital to build a toll bridge on the Trinity River near Palestine. Likewise, the Clarksville Chamber of Commerce pledged land for approach roads if the Austin Bridge Company of Dallas would build a toll bridge across the Red River to connect Clarksville with southeastern Oklahoma.[34]

Oklahomans got an idea of how profitable toll bridges could be when the *Tulsa World* published revenues and ownership of that state's fifteen private bridges. An Oklahoma congressman was earning $9,000 a year on his minority interest in a bridge near Ada. The state Democratic Party chairman owned all or pieces of several bridges, and in 1926 earned $12,000 or more (accompanied by rumors that he might be receiving favorable treatment from the highway department by improving his approach roads). The Purcell-Lexington bridge earned up to 100 percent annually on its valuation of $220,000. Investors in the Red River Bridge Company netted $58,000 in 1926 for their eleven-year-old bridge at the old Colbert crossing. Another newspaper reprinted the *World* editorial, commenting only that the editor "gazes askance at a state that will permit its public to be gypped out of so much money each year."[35]

Public attitudes toward privately owned toll bridges in Texas and Oklahoma soured. Oklahoma had twenty-one private toll bridges within its borders and seventeen interstate bridges, all of them affecting farmers, businesspeople, and recreational travelers from every corner of the state. Some of the toll bridges were already set for condemnation or retirement. An act of the state legislature allowed the Oklahoma Highway Commission to enter into contracts with toll bridge owners with a view toward their eventual purchase. The highway department would grant a bridge franchise to a private operator who would then collect tolls, make a profit, recover the cost of the bridge, and finally turn the bridge over to the state. Some of the older bridges could be condemned as being unsafe for automobiles of the 1920s, but the department had little power to regulate bridges that were already in existence.

In Texas, Ross Sterling remained silent on the matter of privately owned toll bridges. He was fighting other battles: lobbying for another increase in the gasoline tax and trying to build statewide support for an unprecedented $300 million road improvement bond issue. Also, the chairman of the Texas Highway Commission was a hidebound capitalist, a free enterprise man, and he had no desire to see the state compete with a well-run local business. The bridge owner invests his money and takes the risk; therefore, the bridge owner should be entitled to a profit on his invested capital. That was a sentiment Sterling could endorse, but he was wise enough to avoid expressing that view publicly.

Few Texans would agree with Sterling. While they may not have been feeling the same day-to-day toll tax pinch that the Oklahomans felt, the Texans were just as aggravated.

Texas had fewer toll bridges within its borders, and many of those could be condemned, bypassed, or acquired by local authorities. (For example, county officials bought out the owner of a toll bridge over the Pease River between Crowell and Quanah, and then reopened it as a free bridge in 1927.) International toll bridges over the Rio Grande to Mexico were co-owned by the Mexican government; toll rates were minimal and subject to long-standing federal treaties. Of the seventeen Red River toll bridges into Oklahoma, only four were on designated state or federal highways. One of these—the Wichita Falls-to-Lawton bridge on US Hwy. 277—had been rebuilt as a free bridge in 1927. Two of the three remaining Red River toll bridges—the structure on US Hwy. 81 between Ringgold and Terral, and the bridge on US Hwy. 77 between Gainesville and Ardmore—were light and narrow, generally unsuited for modern traffic. Their owners would do little to resist the opening of free bridges at those crossings.[36]

But it was the final bridge, the span on US Hwy. 75 between Denison and Durant, that would lead to an armed standoff between Texas and Oklahoma.

As much as politicians and citizens south and north of the Red River griped about toll bridge charges, it was a former haberdasher and traveling salesman who did something about it.

In 1927, Jacob Jackson "Jake" Loy was thirty-nine years old and single, and had no appreciable political experience, when he won a seat in the Texas legislature from his hometown of Denison. Loy had come to Texas from Tennessee for a railroad job, but ended up employed for several years by Denison's largest dry goods merchant. He left the retail trade to sell work clothing for a Sherman manufacturer, traveling across North Texas and eastern Oklahoma by automobile, bouncing over the rutted roads from town to town, showing off his line of Pool's "Swetpruf" overalls, work pants, and uniforms. A warm smile, a firm handshake, and a sincere interest in his customers' business earned him friends throughout the district. Loy spent a lot of time in his automobile crisscrossing his sales territory. Hours spent on the dirt, gravel, oiled, and asphalted roads every day made him a strong advocate of the good roads movement. Like others, Jake Loy paid his toll each time he crossed the Red River over the Denison bridge and, like others, he came to resent it. He resented it even more when he found that a recent revision to the Federal Road Aid Act prohibited the use of federal funds to improve roads leading to private toll bridges. If the highway for two miles on either side of the toll bridge was to be hard surfaced, Loy learned, he and other county taxpayers would have to bear the full cost.[37]

In the summer of 1926, Jake Loy campaigned for a seat in the state legislature against a wet-behind-the-ears attorney who was on the wrong side of the Prohibition issue. Loy won easily and was sworn in as a member of the Texas House of Representatives on January 11, 1927.

The freshman legislator chalked up two major accomplishments in his first five months in the legislature: he eloped with the engrossing clerk of the House, and he wrote and bulldogged enactment of House Bill 379—the Texas Free Bridge Law.[38]

In fewer than two hundred words, Loy's bill authorized the Texas Highway Commission to allot highway funds for the purchase or construction of interstate bridges over boundary streams. Previously, state courts barred the commission from contracting for work across state boundaries. The only significant limitation in the Texas Free Bridge Law was that Texas could pay no more than *half* the costs of the bridge. To buy the existing bridge or to build a new, toll-free bridge over the Red River would require an adjoining state to pitch in the other half of the cost.[39]

Oklahoma was ready to do just that. Shortly after passage of the Free Bridge Law a delegation from the Oklahoma State Highway Commission visited Jake Loy in Denison, pledging that it would cooperate with Texas in taking over the toll bridges between the two states as soon as funds would permit.[40]

And plenty of funds were on their way. The US government, the states, and the counties would spend an extraordinary $1 trillion on road construction, maintenance, and bridge building in 1927, according to the US

Bureau of Public Roads. Texas had accumulated $7 million of its own, with a like amount coming from the feds. Oklahoma expected to put more than $6 million under contract. There would be plenty of money to buy or build toll-free bridges over the Red River.[41]

Thanks to Jake Loy's bill, there was a chance that the old Colbert ferry crossing—the Denison toll bridge—could be toll free for the first time in almost seventy-five years.

Most local residents were willing to wait for the states to act (although one maverick car dealer in Denison printed hundreds of blue stickers saying "Denison Wants Free Bridges!" and hired schoolboys to paste them on walls throughout the county). Some of those who lived close to the river couldn't wait for the states to act, however. Shortly after Loy's meeting with the Oklahomans, an arsonist set fire to the Devol, Oklahoma, toll bridge upriver from Denison. The midnight fire was discovered and extinguished by bridge attendants, but the firebug returned hours later to set a new blaze and the toll bridge was nothing but char and ashes by daybreak. Several months later, an early morning fire broke out on an eighty-foot section of wooden flooring of the toll bridge between Gainesville and Ardmore. Fire fighters were forced to cut out sections of flooring on both sides of the fire to prevent the whole span from going up in flames. The bridge was closed to traffic for two weeks.[42]

Months dragged by, and there was no move to buy or build bridges over the Red River. The Texas Highway Commission belatedly discovered (with a little help from the state attorney general) that the Free Bridge Law effectively handcuffed Texas to Oklahoma—and as yet Oklahoma had no provision for acquiring interstate bridges. Texas couldn't spend a penny at the Red River without Oklahoma matching it, and the Oklahoma legislature hadn't approved so much as a penny to spend there.

Worse still, directors of the Red River Bridge Company of Denison had quietly informed highway commissioners from both states that the Denison bridge was not for sale. If the states tried to condemn the bridge, bridge company directors warned, they had better look into the franchises granted by the Chickasaw Nation, the state of Texas, and the US Congress.[43]

The states had a strategy, however. It involved a bill introduced by a one-term US senator from Texas, passed by Congress, and signed into law by Calvin Coolidge in 1927. Essentially, the Mayfield Bill decreed that the Red River as it flows between Texas and Oklahoma—that braided ocher stream, mostly mud, and said to be so shallow that an armadillo could walk across it—would thereafter be a "navigable waterway" and subject to oversight by the US Department of War.[44]

There is a vast body of law—often contradictory and usually confusing—related to navigable waterways. The laws came about to assure that our nation's waterborne transportation was treated consistently, regardless of which state may be adjacent to the waterway. The federal government dictated the placement of lighthouses, the operation of harbors, the height of bridges over river channels, the use of piers, and whatever else was necessary to keep water traffic and navigable waterways moving and safe. Navigable waterways were placed under the jurisdiction of the Corps of Engineers, part of the War Department. Asked to determine what constituted a "navigable waterway," the US Supreme Court decided that "navigable in fact is navigable in law." That is, if commercial boats and ships can travel on it, it is probably a navigable waterway.[45]

The Red River between Oklahoma and Texas was never commercially navigable. From time to time a lone steamboat might chug its way as far up the river as Denison, but come summertime the boat would be mired in mud. The river was not "navigable in fact," and so not "navigable in law."

That is, until the US Congress declared it otherwise.

With a snap of congressional fingers, the Red River between Texas and Oklahoma became a navigable waterway, giving the War Department power to regulate tolls on all bridges constructed across it. As Texas senator Earle B. Mayfield gloated during his reelection campaign, "The public that uses these bridges can now be given relief from the extortionate charges which hitherto they have been assessed."[46]

Texas and Oklahoma could now employ the Free Bridge Law—granting Texas the right to buy or build a toll-free bridge over the Red River—and the Mayfield Bill—giving the Corps of Engineers the ability to regulate tolls over the interstate bridges—as the carrot and the stick to open up toll-free crossings over the Red River.

Oklahoma was showing signs of passing a bridge bill of its own. Angry editorials helped animate legislative candidates for the 1928 legislative session, and the politicians weren't long in expressing their constituents' dissatisfaction. The *Daily Oklahoman* asked every candidate in the state to declare his position and found that every office seeker opposed private toll bridges. "I do not believe that a toll bridge has any place in a free country and among a free people," one candidate declared flatly. Another candidate for state representative, an ardent supporter of better roads, agreed: "A toll bridge should be and is obsolete with the progress of today."[47]

"Is it worth four times as much to cross a Red River bridge between Oklahoma and Texas than it is to go over the international bridge at Laredo?"[48]

On January 6, 1928, the Texas attorney general's office announced plans to discuss just that question with their counterparts from Oklahoma and members of the two states' highway departments. Their purpose, an assistant attorney general from Texas said, was "to force a material reduction in the tolls collected by the Red River bridges." The group enlisted a committee of engineers from the two states and the Corps of Engineers who would appraise the bridges, determining their value and return on investment for the owners.[49]

The Red River Bridge Company, owner of the Denison toll bridge, began searching for the shrewdest lawyers it could find.

CHAPTER 4

The First Casualty

THE VOICES OF motorists, automobile groups, legislators, and news editors against toll bridges rose in volume and stridency throughout the 1920s, leading editorialists of the *Washington Post* to proclaim that "there should be no place on the publicly owned highways for privately owned bridges." The existence of toll bridges contradicts "the whole principle upon which the nation and the states have based their highway construction programs."[1]

The highway departments of Oklahoma and Texas were committed to ridding their states of toll crossings. The Oklahoma legislature struggled to pass funding legislation similar to Jake Loy's Free Bridge Bill, but kept getting bogged down by a faction that wanted to buy out the state's intrastate toll bridges first. Texas highway commissioner Cone Johnson made it clear he "would never vote a dollar of the people's money to build a state highway to a privately owned toll bridge."[2]

The interstate toll bridge causing the most public clamor was the Red River toll bridge between Denison and Durant. Texas and Oklahoma wanted to eliminate tolls at that highway crossing, but didn't seem to know quite how.

Throughout the fall and winter of 1928, the states strategized about how to force the Red River Bridge Company owners to sell their toll bridge (or how to muscle them out of business altogether). Their initial strategy depended on intervention by the US secretary of war, who had the power to dictate tolls for bridges over navigable waters.

After a few perfunctory hearings (and with a little prodding from the states' attorneys general), Secretary of War Dwight F. Davis sent notice to the bridge companies on December 31, 1928, that current tolls to cross the Red River were exorbitant and must be reduced. The order required the Red River Bridge Company to cut toll rates in half immediately. Directors of the Red River Bridge Company refused, seeking and winning a temporary injunction against the secretary of war while continuing to collect tolls at the higher rate.[3]

Failing to starve the toll bridge owners out of business through reduced rates (and with the Red River Bridge Company refusing to sell), Texas and Oklahoma next considered building their own free bridge parallel to the toll bridge. Texas had already passed Jake Loy's Free Bridge Bill, authorizing the state to pay half the cost of an interstate bridge, but Oklahoma lacked a similar authority. Though most lawmakers supported the action in principle, no Oklahoma bridge bill was yet in sight; it remained locked in committee through the fall and winter.[4]

By the beginning of 1929 it was apparent to all parties that the matter of the Red River toll bridge was headed to court: Oklahoma and Texas to open up a free crossing; Red River Bridge Company to protect its business and the value of its franchise. The year 1929 would mark the beginning of eighteen months of legal wrangling in Texas and Oklahoma state courts, federal district courts, and before the Supreme Court of the United States.

The first casualty of what would later become known as the Red River Bridge War arose from that litigation and occurred on Saturday, April 13, 1929, in a Grayson County district courtroom in Sherman, Texas.

Attorney General Claude Pollard of Texas, Attorney General Edwin Dabney of Oklahoma, and a handful of their assistants held a secret planning meeting in a Sherman hotel room on April 5, 1929. The meeting didn't stay secret for very long; the following day, Assistant Attorney General Rice M. Tilley revealed to reporters that the state of Texas would be filing suit against the Red River Bridge Company, possibly before the end of the week.[5]

The eager assistant wasted no time. On Tuesday, April 9, Tilley filed suit on behalf of the state of Texas against the directors of the Red River Bridge Company in the fifteenth district court in Sherman. The suit asked that the toll bridge at the old Colbert's ferry crossing between Denison and Durant be declared a free bridge on the grounds that the Texas franchise under which it operated had expired and was therefore null and void. Tilley asked the judge to restrain the bridge owners from collecting tolls, citing grounds of fraud against the public. Tilley's petition went a step further: in the event the court finds that the franchise is valid, the lawsuit continued, the court should require the bridge owners to lower tolls in accordance with the order issued by the secretary of war earlier that year.

The judge denied the immediate injunction but scheduled a hearing for the following Friday.[6]

Tilley's filing was just the first of a double-barreled blast by Texas and Oklahoma. Two days later, on Thursday, April 11, an Oklahoma assistant attorney general filed a similar suit in the state district court in Durant. The state of Oklahoma contended that the bridge company was operating in violation of Oklahoma statutes, not having acquired a franchise from that state. The toll bridge was interfering with the right of the public to free use of the bridge, according to the attorney general, and he asked for an immediate restraining order.[7]

After a short hearing in Durant that afternoon, Oklahoma district judge Porter Newman agreed: "It is hereby ordered and adjudged that the Red River Bridge Company of Texas, its officers, agents, and employees, are hereby temporarily enjoined from barring and obstructing the bridge, and from charging and collecting tolls."

Judge Newman dispatched the county sheriff, accompanied by several of the bridge directors, to tell the toll taker to cease collecting tolls. By sundown—for the first time in eighty-two years—any pedestrian, bicyclist, wagon, horseman, bus, truck, automobile, or motorcycle could cross the Red River at the old Colbert's ferry site without paying a cent.

The bridge didn't remain toll free for long. The next morning, the owners arranged to move the tollbooth, the toll taker, and his chain from the northern (Oklahoma) approach road to the southern (Texas) approach road. Out of Oklahoma's jurisdiction, the toll taker commenced collecting tolls as usual.[8]

These early skirmishes accomplished little; the big showdown would occur two days later, on Saturday, April 13, in the district courtroom in Sherman, Texas.

Sherman is ten miles south of Denison and, like Denison, sits astride the King of Trails Highway. But where Denison is the Gate City—drawing its trade from the surrounding countryside, tourists passing through, and the rail passengers who arrive and depart from Denison's three depots—Sherman is the seat of Grayson County, benefiting from the attorneys, landmen, contractors, and others who come to the courthouse daily. Saturday is Market Day in Denison, the sidewalks, stores, and wagon yards crowded like a carnival midway; Sherman is a Monday-through-Friday town, with only the largest retailers, the hotels, and the occasional chili parlor open for business on Saturday.

On this second Saturday of April, however, autos filled the spaces on all four sides of Sherman's courthouse square and people streamed across the lawn, past the Confederate veterans' statue, up the steps, and through the doors of the courthouse for a rare weekend session. There was standing room only—and precious little of that—in Judge Roger M. Carter's second-floor courtroom by eight-thirty that morning to watch the fireworks promised by *State of Texas v. Red River Bridge Company of Texas.* Representative Jake Loy, author of the Texas Free Bridge Law, was in the gallery, along with Bryan county commissioner "Doc" Diffey from Colbert. Several of the major toll bridge shareholders were in the audience: brothers Walton and Clyde Findley, and Miss Margaret Kelley, vice president of the National Bank of Denison. Two assistant attorneys general from Oklahoma, representing Attorney General Edwin Dabney, sat with Oklahoma highway commissioner S. C. Boswell (who was also a bridge company shareholder). Claud Easterly of the *Denison Herald* covered the hearing, along with reporters from the big-city newspapers—Dallas, Oklahoma City, Fort Worth, Tulsa. The reporters were allowed to sit in the jury box, but no photographers were allowed in the courtroom.[9]

Representing the state, Rice M. Tilley spread his papers out on the plaintiff's table, leaving room for local counsel. Tilley was a young man, barely thirty years old and only seven years out of law school, but he had built a fearsome reputation as an interpreter (and enforcer) of the new body of law relating to highways and auto travel. Tilley was rail thin, dark eyed, and serious; and he wore an intimidating expression in the courtroom: the look of a cantankerous dog unafraid, even eager, to bite (and would probably enjoy it when he did).

Tilley was the polar opposite of his jovial and portly opponent, the Honorable US Senator Joseph W. Bailey.

Joe Bailey was the personality the crowd had come to see, and when he walked into the courtroom his presence hushed all chatter, as if a dishtowel had been thrown over a birdcage. Bailey was an outsized man—six feet tall and close to three hundred pounds—with an outsized political and legal reputation in Texas and among conservative Democrats nationwide.[10]

Born during the Civil War, Bailey earned his law degree; then, at age twenty-eight, he was elected to the US House of Representatives. He was a remarkable orator, and it was said that "his tones lingered in the chamber like the echo of chimes in a cathedral." [11] Bailey served five terms in the House—two of those as House majority leader, the youngest in its history—before being elected by the Texas legislature to the US Senate in 1901. Bailey was known by colleagues in Washington for "his forthright declarations of opinion and his penchant for trying to reinforce these declarations with his fists." He resigned after two terms rather than face censure for some lobbying irregularities (and for punching out a senator who heckled him). Bailey opened a law practice in Dallas with his son, specializing in transportation and oil law, while continuing to dabble in elective politics. In the courtroom, his oratorical flourishes and formal manner were reminiscent of an earlier century, but he was known as an exceedingly clever legal tactician with a near-photographic memory of constitutional case law. So far, he had helped the Red River Bridge Company dodge every bit of legal cookware the states of Texas and Oklahoma threw at it.

Bailey walked down the center aisle toward the defendant's table, smiling a greeting at those he knew and shaking hands as he went. He was sixty-six years old and wore a longer-than-fashionable Prince Albert suit coat that hung loosely on him with a hint of Roman drapery, his salt-and-pepper hair brushed straight back off his face. Even as a young man, his face had been fleshy, with thick lips, sagging cheeks, and a prominent brow. At age sixty-six, those features were exaggerated, and several rows of jowly flesh wobbled and jiggled beneath his chin. The old politician sighed with an audible wheeze as he lowered himself into a chair at counsel's table.

Sitting next to Bailey in the courtroom was Joseph W. Bailey Jr., his father's law partner and biggest fan. As a child, Junior sat perched on his father's shoulders as Senior paraded him around the capitol, introducing him to fellow lawmakers. The boy learned to mimic his father's oratorical gestures, which earned him laughs from Senior's friends and the nickname "Little Joe." Little Joe earned headlines of his own when Big Joe took his son to a meeting at the White House. Little Joe refused to sit on Pres. William McKinley's knee because he was a Republican and had to be carried, wailing, from the room. Now thirty-six years old (and still proud of his nickname), Little Joe practiced law with his father, preparing briefs and managing their office while Senior most often appeared in court.[12]

At seven minutes past nine o'clock, Judge Carter gaveled court into session.

"This matter goes far beyond the issue of toll collection," Rice M. Tilley told the court during his opening remarks. In fact, the privately owned toll bridge had become a painful carbuncle on the rear end of a two-year-long effort by Texas and Oklahoma to complete a network of interstate highway connections. A river-crossing franchise granted seventy years before was blocking the two states' plans to complete their section of the 1,700-mile King of Trails Highway intended to carry tourists, commercial travelers, and goods from the port of Galveston, Texas, to Winnipeg, Canada. Worse yet, the states were losing hundreds of thousands of federal road improvement dollars every year as long as an "extortionate" toll bridge was allowed to block the interstate highway. Politicians and motorists demanded a solution.

Once again, Tilley proposed that the court declare null and void the exclusive franchise under which the narrow, two-lane toll bridge operated, opening the bridge to free passage. Failing that, Judge Carter should require the bridge company to lower its tolls in accordance with the secretary of war's order, an order still under consideration in a Washington, DC, federal court.

Bailey waited politely for Tilley to take his seat before rising to spring one of his tactical surprises. With a self-confident smile (and after a flowery lecture to the court on constitutional law), the flamboyant attorney asked the district judge to transfer jurisdiction of the entire case to a federal court, where the suit might be resolved more quickly as a matter of interstate commerce.

Tilley did not want to lose control of the case. The young state's attorney knew the case, knew the law, and knew he would ultimately prevail. On the other hand, Tilley recognized that the crafty old attorney might be offering him a plum. A move to federal court would consolidate and simplify the cases against the Red River Bridge Company, possibly accelerating a decision. But Texas and Oklahoma would lose their seat behind the steering wheel to federal attorneys, who might or might not proceed in the two states' best interests.

The decision about how to respond to Bailey's surprise move was above Tilley's pay grade, and he asked for a brief recess to consult with the attorney general in Austin.

As Tilley placed frantic phone calls, Bailey was his usual voluble self in the courthouse lobby. The big man held forth on general matters of politics and law while visiting with friends from the spectators' gallery. Little Joe, as usual, stood behind his father as the senior Bailey answered a reporter's questions, joking about his recent decision to represent the Association Opposed to the Prohibition Amendment in a state known for its "dry" preferences. Bailey wasn't afraid of unpopular cases.

Back in the courtroom, with all parties seated at the counsel tables, Tilley shuffled papers as he prepared to respond to Bailey's unexpected motion. In a surprising breach of etiquette, Bailey rose suddenly and opened his mouth as if to speak.

All he managed was a choking gurgle before collapsing into the chair next to his son, dead of a stroke before his face hit the counsel's table.

The Hon. Joseph W. Bailey Sr. was the first casualty of the Red River Bridge War.

At its core, the Red River bridge dispute should have been a simple matter at law: the owners of the toll bridge, with exclusive licenses from the US Congress, the state of Texas, and the Chickasaw Nation, had borne the costs of building and maintaining a for-profit bridge. When Texas and Oklahoma attempted to make their bridge a free one, it could be said they were depriving the toll bridge owners of their livelihood, and the states should compensate the license holders for their loss.

The only issue on the table, then, ought to have been to determine a reasonable amount of loss. But when a gaggle of lawyers representing two states, various state departments, and even the federal government gather around the table, "reasonable" can lose its seat.

Former senator Joseph W. Bailey may have been moving all parties toward a reasonable solution, but when he keeled over in the Sherman courtroom in April 1929, the parties scattered to protect their manifold interests.[13]

Bailey's untimely courtroom death entangled the states of Texas and Oklahoma, the Red River Bridge Company, and the federal government in a vicious legal snarl, a yearlong running gunfight that would generate front-page headlines across both states and result in the Red River Bridge War.

Just five days after Joseph W. Bailey's funeral service, Little Joe—determined to complete what his father started—filed papers with federal courts in Texas and Oklahoma, alleging "a malicious conspiracy or understanding among highway officials, attorneys general, and other state officials of Texas and

Oklahoma." The bridge company's suit asked for the protection of the federal government, and requested that the court appoint a receiver to watch over the interests of the bridge company.[14]

Texas Assistant Attorney General Tilley responded by refiling his suit in the state court at Sherman, again asking that Judge Carter bring the case back to state court. Judge Carter did grant a temporary restraining order prohibiting the bridge company from collecting tolls. The bridge company simply moved its toll taker back to the Oklahoma end of the bridge, where Oklahoma's earlier order had expired.[15]

A federal judge in Muskogee appointed J. R. Handy, a Denison insurance man and vice president of the Red River Bridge Company, as an operating receiver. Handy was sworn to preserve and maintain the assets of the Red River Bridge Company pending the conclusion of litigation.[16]

Meanwhile, Judge Carter in Sherman granted the injunction that Rice Tilley had requested. There was one vital exception: as a state court judge, Judge Carter could declare the earlier bridge franchises void and prohibit the further collection of tolls, but he couldn't prohibit toll collection by receiver J. R. Handy, who was doing so *under federal direction.* The status quo prevailed, and a resolute Rice Tilley was forced to look for more ammunition.[17]

Within days Tilley filed suit against Handy and the Red River Bridge Company for fraud, alleging the company had been collecting tolls illegally, without a valid franchise, since 1914. He asked the judge for forfeiture of $300 a day in tolls and an equal amount in damages, a total of about $3.2 million.[18]

The Red River Bridge Company and its legal team survived these legal shoot-outs for most of 1929, but their luck ran out during the very last moments of a special legislative session at the state capitol in Oklahoma City.

The strategy of the state of Texas had been to litigate the Red River Bridge Company out of business (or at least to force a sale). Under Jake Loy's bridge bill, the state was authorized to buy the existing bridge, but was prohibited from paying for more than half a bridge. Oklahoma had not yet passed legislation allowing it to pay for the other half of the bridge. That was about to change.

Oklahoma's legislative session had been hijacked by a trial of impeachment against the current governor and had failed to pass a number of important bills. The new governor called for a special session with clear instructions to wrap up the business of the toll bridges over the Red River. In the waning moments of the special session, both houses passed a hastily written bill giving the Oklahoma Highway Department authority to pay half the total costs of building toll-free bridges on US highway routes across the Red River.[19]

With both states now having the authority to build a bridge of their own, Red River Bridge Company owners came to the bargaining table on October 22, 1929. At a meeting in a Sherman hotel room with representatives of both state highway departments (and the ubiquitous Rice Tilley), Joe Bailey Jr. and J. R. Handy offered to *give* their bridge, along with all franchise rights, and properties owned by the bridge company to the states if the states would allow them to collect tolls unmolested for an additional seven to ten years. Bailey calculated that Oklahoma and Texas would save as much as $300,000 each on construction fees and litigation while allowing the bridge company to cash out their shareholders. Everybody wins.[20]

The state highway men didn't see it that way; their legislatures had given them the money and the mandate to *build* brand-new bridges, not buy rickety old ones. After brief consultations with their state capitals, the highway men rejected the offer out of hand. A week later, the state highway departments placed advertisements for bids on three bridges across the Red River. The existing Terral-to-Ringold and Gainesville-to-Ardmore bridges were worth little more than salvage, and the states could buy them out for pennies; the Denison-to-Durant bridge was no longer a consideration. After all, even if the toll bridge remained in operation, who would choose to pay a toll when a brand-new free bridge was available? Bids for the new bridge were due by November 25, 1929.[21]

The one person who should have been most vocal about Red River highways and toll bridges was the chairman of the Texas Highway Commission, Ross S. Sterling. However, during two years of strategizing, lobbying, and litigating the Colbert bridge issue by the state of Texas and his own highway department, Sterling never commented publicly. Perhaps it was because he was, at heart, a free-market capitalist who found chastising the profitable little toll company distasteful. Perhaps he didn't want to call attention to the business and personal relationships he shared with owners and directors of some of the Red River bridges. (H. W. Stark, Gainesville pharmaceutical distributor and part owner of the toll bridge there, nominated Sterling for a seat on the Texas Christian University board of trustees and would provide financial support to Sterling's later political campaigns; Margaret Kelley, vice president of the National Bank of Denison and an officer of the Red River Bridge Company, was a professional protégé of Sterling's sister, who helped establish Kelley as one of Texas' foremost women bankers.)[22]

Perhaps it was because Ross Sterling was considering a run for governor.

Sterling had never run for public office, but he was an ambitious man, proud of his accomplishments with the Texas Highway Department. By the fall of 1929—after three-and-a-half years chairing the commission—he had signed more contracts for more money and provided more jobs than had any man in the history of the state. Texans were now driving on more miles of hard-surfaced, all-weather highway than were drivers in any other state in the nation, and Sterling had accomplished all of that with no favoritism and not even a whiff of scandal.[23]

Meanwhile, the state's economy was flourishing. Tax revenues from petroleum exploration and production were enriching the state. Texas cotton was selling at $19.75 a pound, down a bit since the end of the Great War, but farmers were buying more land for production and financing it with low-interest mortgages available through local banks. Chrysler had just introduced its new Dodge Brothers Six Coupe; at $595, it was an automobile any workingman of the day could afford. A confident middle class was demanding—and buying—luxuries; even small-town newspapers carried ads for European millinery, nylon stockings, and gold pen sets. There were on the horizon only the faintest puffs of smoke from the prairie fire that would be known as the Great Depression. Sterling ignored these warning signs; he saw ahead an illimitable vista of prosperity.[24]

Texas was becoming big business and, Sterling reckoned, needed a big businessman to manage it.

Governor Moody, the man who appointed Sterling to the highway commission, hadn't made his reelection plans clear, so Sterling kept his own counsel for the time being. But he used his trips around the state as chairman of the Texas Highway Commission to meet political heavyweights and test the waters. It was on one of those trips in late October 1929 that he picked up a newspaper in the coffee shop of the Nueces Hotel in Corpus Christi to read about financial panic on Wall Street. "I did not dream and could not have believed," he said later, "that the morning's news marked the beginning of a financial hurricane that was to wipe out my entire fortune."[25]

Meanwhile, he remained mum about problems surrounding the Red River bridges. If he were to run for governor, he could take credit for the three new free bridges without wearing the stain of how they were acquired.

The Red River Bridge Company would not go gently, however.

As engineers for the construction contractor began surveying likely sites for the new bridge, they discovered that the Red River Bridge Company owned land on both sides of the river, contiguous parcels that prevented access to the river. Some of the parcels dated back to 1915 and plans for the aborted interurban power station; others had been acquired in the previous year. To obtain the land, Texas and Oklahoma would have to institute separate condemnation proceedings in separate state courts, proceedings that would likely be removed to federal court and subject to still more delay.[26]

In March 1930, the Oklahoma attorney general delivered an opinion that the Chickasaw Nation had no authority to grant exclusive crossing rights to Frank Colbert in 1876. He issued the opinion fearing that the Chickasaw legislature might become involved in blocking construction of the new toll-free bridge. A riled-up tribal legislature did just that, filing suit against Oklahoma in federal court.[27]

These lawsuits (and other bridge-related actions) came together in a federal courtroom in Muskogee in May. Discussions became heated, and a frustrated federal judge, R. L. Williams—who had some twelve active toll bridge cases piled up on his docket—ejected attorneys Rice Tilley and W. C. Lewis from his courtroom and suspended the proceedings.[28]

Several court-ordered attempts at negotiation went nowhere, and on June 11 Judge Williams's clerk received still another filing, this time from the Red River Bridge Company seeking to restrain any bridge contractor from beginning work on a new bridge. The bridge company threatened personal suit against any contractor's employee who as much as put shovel to dirt.[29]

By spring 1930, the whole mess had devolved into a bloody knife fight, with all parties worn out but still slashing. That's when Ross Sterling spoke up.

"I have decided to enter the Democratic primary as a candidate for governor," Sterling announced on May 30, 1930. He had been encouraged, he said, to "administer the state's business as [I] have administered the highway department."[30]

Sterling jumped into an eleven-candidate race as a definite longshot. Voters had never chosen a businessman to be governor of Texas, an office previously held sacred for politicians, usually trained only in law. Sterling's campaign was unique in another way, one observer said: men with almost no political experience dominated his organization, and that included Ross Sterling himself.

Texans yearned for a leader who could help them make sense of what was happening with the state's economy. In the six months since the Wall Street crash, Texas was producing more oil than ever before, but because prices had dropped through the floor, the state's tax revenues were drying up. Because oil prices were down, refiners and shippers were laying off workers. Cotton was down to $15.10 a pound—off 25 percent since the previous October—and farmers who had taken mortgages to buy additional acreage were facing foreclosure or worse if prices didn't rise by market time. And then there were the bank failures: banks in Aubrey, Post, Ennis, Merkel, Jefferson, Henderson, Brownwood, and two dozen more in the first five months of 1930 alone. Thousands of families lost their savings, small-town merchants lost their lines of credit, churches lost their building funds, and small towns lost their collected tax money; the impact was as devastating as it was inexplicable. Voters came to believe that Ross Sterling—oilman, farmer, banker, businessman—could deal with these problems.[31]

From the first, Sterling's physical appearance drew comment. At 250 pounds on a six-foot-three frame, Ross Sterling was no heavyweight in the class of, say, Grover Cleveland or Chester A. Arthur, but Texas politics require a certain amount of self-deprecating humor. When an opponent disparaged him as "that fat boy from Houston," Sterling adopted the name and began promising every campaign audience that "this fat boy from Houston will work for you."[32]

Despite being more comfortable speaking to men in suits and ties than to farm families in overalls and feed sack dresses, Sterling was a good campaigner in Texas' rural areas. In West Texas especially, he said, "when the cowpunchers came to see me they expected to find a city man who smelled of perfume and looked soft of muscle." When they made a man-to-man survey and started talking cattle, the Texans recognized him as one of their own. Sterling grew up with the language of country people, but it embarrassed him when he used "I seen," "I done," or "I taken" in his formal speeches. A supporter who shared the platform with Sterling in Terrell noted his discomfort and told the crowd: "Mr. Sterling may say, 'I done,' but when he says that, you can bet your boots he has done it. And when he says, 'I taken,' you can be dead sure that he has taken only that which belongs to him."[33]

Sterling's straightforward speeches and his reputation for executive competence and honesty were taking him to the top of the eleven-man primary field, but his public reputation was threatened by the continuing clown carnival of litigation between his highway commission and the Red River Bridge Company. Throughout the primary campaign, state newspapers reported on the Red River bridge squabble, growing more frustrated every day with the lack of progress toward a new, toll-free bridge over the Red River. Sterling's support—already soft in North Texas—was in danger of collapsing altogether.

On Monday morning, June 23, a month before the Democratic primary, principals in the Red River bridge litigation arrived at the Texas Highway Department for a private meeting. Attorney Joe Bailey Jr. and his client, J. R. Handy, receiver of the Red River Bridge Company, had arrived in Austin by auto the previous evening. Oklahoma Highway Commission chairman Lew H. Wentz and his chief highway engineer, A. R. Losh, had taken the early train from Oklahoma City. Oklahoma Attorney General J. Berry King and his first assistant, W. C. Lewis, were in attendance, along with their Texas counterparts, Claude Pollard and assistant Rice M. Tilley. Texas highway commissioner Cone Johnson and state engineer Gibb Gilchrist also participated. With all parties assembled, the dark oak doors to the private conference room closed, and the meeting began.

There was no public agenda for these meetings and no list of attendees. The capital press noted the gathering, but the only comments they collected from participants were vague and intended to misdirect. Ross Sterling, chairman of the Texas Highway Commission and a gubernatorial candidate, took those two days off from his campaign, but his precise whereabouts are unrecorded.

Tight-lipped participants gave little clue about what occurred at the meeting.[34]

The Oklahomans abruptly packed their briefcases at the end of the first day and returned home, chairman Wentz later expressing their inability to find common ground. Bridge company representatives and Texas highway people, however, met for a second day and part of a third before leaving quietly and returning to their offices. Reporters noted that within a week following the meeting, attorneys for both states made a surprising turnaround, joining bridge company attorneys in asking courts to delay or dismiss Red River bridge cases already in progress.

The public was not aware (and wouldn't know for more than a year) that the outcome of the private meeting was a set of contracts. Little Joe Bailey and other bridge company attorneys struck a secret deal. Rice Tilley reviewed the contract drafts and, after minor clarifications and corrections, the president of the Red River Bridge Company of Texas signed contracts with the Texas Highway Commission, contracts intended to end all litigation and clear the way for completion of the new toll-free bridge over the Red River.

Ross "Fat Boy" Sterling's come-from-behind campaign took him to second place in the Democratic primary, garnering enough votes to enter the runoff election. Earning the endorsements of his defeated opponents, Sterling won the nomination with 56 percent of the vote.

Sterling won the Democratic Party nomination due, in no small part, to his reputation for competence in running the Texas Highway Department. He lifted Texas motorists out of the mud with better roads and a highway system rivaling any in the country; he was in the process of clearing away antiquated toll bridges and opening modern free bridges across the Red River. Continuing litigation and delays would reflect poorly on Sterling and decrease his chances in the race.

But someone had brokered a bridge deal in time for the general election: someone with a personality strong enough to muscle all interested parties into the same room; someone with the business negotiating skills necessary to gain for every party what each needed to end the legal warfare; someone willing to strike a bargain and seal it with an honest handshake.

Based on later evidence, it is reasonable to conclude that Ross Sterling directed the settlement, and it was the signature of Ross S. Sterling—chairman of the Texas Highway Commission, clean-as-a-whistle businessman, and soon-to-be-governor of Texas—which would validate the contracts that were never intended to see the light of day.[35]

Under the terms of the agreements that were signed on July 24 by Sterling, Commissioner Cone Johnson, and the Red River Bridge Company, the company promised to withdraw all objections to the Texas-Oklahoma free bridge and dismiss all litigation. Red River Bridge Company would do nothing to interfere with the construction or operation of the new bridge; the company would give clear title to any property needed for southern approach roads to the Texas Highway Department and would convey easements to Oklahoma for approach roads from the north.

In return, the toll bridge would continue to operate and collect tolls until the opening of the free bridge—a minimum of fourteen months—and the state of Texas would pay $60,000 to the Red River Bridge Company when the free bridge opened. If the new bridge opened sooner than fourteen months from the contract date, Texas would pay the Red River Bridge Company an additional $10,000 a month for each month of toll collection missed. The state also agreed to become a party to the bridge company's long-pending lawsuit against the US Department of War, which had been trying to force a reduction of toll rates.[36]

There was no formal announcement of the agreement, but on October 15, 1930, the Texas attorney general's office reversed its long-standing position by opposing the War Department's insistence on lowering bridge tolls across the Red River. Meeting with the Corps of Engineers district officer in Dallas (and over Oklahoma's objections), Assistant Attorney General Rice M. Tilley joined with the bridge company in opposing the decrease.[37]

As Sterling campaigned across the state, construction resumed on the bridge north of Denison. Surveyors surveyed, road graders graded, drillers drilled bridge piers, and ugly headlines about lawsuits and delays disappeared from newspapers.[38]

On November 4, 1930, Ross "Fat Boy" Sterling was elected governor of Texas.

CHAPTER 5

Cheese and Sody Crackers

IN 1929, as Oklahoma, Texas, and the Red River Bridge Company fought it out in courtrooms and Ross Sterling considered his run for the statehouse, a fifty-nine-year-old small-town lawyer/politician was trying to raise the fare for a return to his adopted home. His dream of an agricultural paradise had melted away in the tropical heat.

"'Alfalfa' Bill Sells His Eden in Bolivia and Plans Return to Oklahoma."[1]

New York, Washington, Chicago, Los Angeles—newspapers across the nation reported on the collapse of a former two-term congressman's dream to build an American agricultural colony in the jungles of Bolivia.[2]

"'Alfalfa Bill' was one of the striking characters in congress a dozen years ago. He was a member at large from Oklahoma, and his tousled shock of wiry hair, his sweeping buck-horn mustache, and his ever-present quid of chewing tobacco did much to give variety and color to the congressional landscape."[3]

Nine years earlier, defeated for a third term in congress and believing his public service career was over, former congressman William H. Murray began laying plans to develop a new agricultural society in South America. From his family farm near Tishomingo, Oklahoma, Murray traveled first to Peru, where he negotiated a land concession with the president. The Peruvian president promised to build an automobile road to the land there, but reneged on his promise.

"Although we had contracts we had to give them up," Murray said. "There was nothing but a mule trail to the lands I had selected."[4]

He next explored the Chaco area of Bolivia, eventually signing a ninety-nine-year lease for forty-two thousand acres (with an option for an additional two hundred thousand acres) of flat, cleared lowlands. Murray returned to

Oklahoma to recruit the twenty-five families required to meet the terms of his lease.[5]

"Colonizing Bolivia is going to take a lot of work," he said, "but so it did in Oklahoma." In his prospectus for colonists, the former lawmaker laid out his rules: no liquor, no gamblers, no slackers. He cautioned those who yearned for bright city lights or did not relish hard manual labor to keep away.[6]

In 1924, Murray cashed out everything he owned in Oklahoma and led eighty-two colonists (including his sons and their families) to new homes in South America. Despite their enthusiasm and Murray's warnings, several families found conditions too primitive, too difficult, and asked Murray for funds to return to Oklahoma. An overlong rainy season, a plague of locusts, and harassment by local Indians prevented the colonists from making profitable crops the first year, and two more families packed up and left. Worse still, a road to the nearest railhead promised by the Bolivian government failed to materialize, making it difficult to get crops to market.

"The rains, lack of roads, lack of markets, and too much jungle wrecked Bill's dream of a tropical Utopia," one newspaper reported.[7]

After five years of backbreaking labor and a loss of hope, Murray had to admit failure: "When I landed in New Orleans on the thirteenth of August, 1929, I had forty-five dollars left, with nothing but our personal wearing material packed in two trunks, and four thousand pounds of my library. No household goods; no furniture of any kind."[8]

Almost sixty years old, W. H. "Alfalfa Bill" Murray would be starting over from scratch once again.[9]

William Henry David Murray was born in 1869 in the Texas community of Toadsuck (near present-day Collinsville) less than forty miles southwest of Frank Colbert's ferry crossing on the Red River. His father ran a gristmill and was a part-time evangelist; his mother died in childbirth when Murray was two years old, leaving three sons. The Murray household never lacked discipline, especially after his widowed father married a woman neighbors called "a religious fanatic."[10] The boy quickly tired of the discipline. Telling his father and stepmother he was going to a nearby community for a church meeting, twelve-year-old Bill Murray walked away from Toadsuck and didn't return for almost fifty years.

The itinerant teenager picked cotton here, hauled lumber there, found work in brickyards and cotton gins, and generally did odd jobs where he could find them. Between crops or during the summer months he attended the local school, but his early education was erratic and incomplete.

It was in his late teens that Murray became serious about a formal education, trading his labor for tuition and board at College Hill Institute northwest of Fort Worth. He became a voracious reader and joined the debate club. He studied history, logic, political oratory, the classics, and many of the political and economic treatises favored by agrarians of the time, eventually earning a bachelor of arts degree and winning a teaching job, where students nicknamed the raw, lanky young man the "Professor."

Little as it was, such learning was rare in unlettered rural Texas. Professor Murray spoke up for the farming interests of his county. He became an activist in the Farmers' Alliance and Democratic Party, exhibiting his talent as a homespun orator; he wrote editorials in support of lower interest rates for farm loans, support for farm prices, better transportation to markets, and lower costs for storage. "If Mr. Murray were not too young," one newspaper editor wrote, "he would be good timber out of which to construct a state senator."[11]

Still in his twenties, Murray chose to swim into the maelstrom of Texas politics. In 1892, he announced his candidacy for the state legislature and was defeated handily in the primary. He won the job of secretary to the state Democratic Party convention and ran the show with a steady hand, due largely to his knowledge of parliamentary procedure. One writer observed that political reversals "seemed only to whet [Murray's] appetite for leadership and recognition."[12]

Murray and his brother started a newspaper in Corsicana in 1894, largely to further Murray's political ambitions. Murray's editorials celebrated the yeoman farmer of the last quarter of the nineteenth century. Above all else, he propounded the extension of the family farm and the preservation of rural America, and he wasn't too dainty to take his fights to the street from time to time. In a second campaign for a seat in the state senate, Murray had the support of farm families, but he was edged out by voters from the more urban sections of the district.

A voracious reader, Murray began devoting his time to the study of law. One night a week, he would meet with a friendly judge who would ply the young man with problems of logic and questions of law. By the time he passed his bar exam in 1896, Murray had grown frustrated with the pettiness of local politics. He and his brother closed the newspaper, settled their debts, and Murray moved to Fort Worth to hang out his shingle. According to one writer, turn-of-the-century Fort Worth was afflicted—infested, even—with an overabundance of lawyers (few of whom were prospering). What little legal work Murray could find involved commercial collections, but he retained plenty of time for reading and study. Typically, he burned the candle at both ends for almost two years, and never seemed to run out of wax.

Money was another matter. The lack of clients, he said, put him on a pie and coffee diet, and his financial condition grew increasingly tenuous. In March 1898—after two unsuccessful political campaigns, a failure in the newspaper business, a stillborn law career, and with just ten dollars in his pocket—Murray gave up his law practice and boarded a train that would take him north across the Red River into Indian Territory.

Not yet thirty years old, W. H. "Bill" Murray would be starting over with nothing.

Murray never gave a definitive reason for moving to Indian Territory when he did. Friends said the decision was made on a coin flip: heads would take him north, tails to South America. Others said he left Texas after rejection by the family of a sweetheart. But Murray's father had moved to Marlow in the Territory several years before, and the son heard enough stories about the Nations to convince him that the area held promise for a hard-working, well-read young man with a law license and a thirst for politics. Murray first arrived in Ardmore where a friendly judge suggested he travel on to nearby Tishomingo. Plenty of opportunity there, he assured the gawky young man.

It was in Tishomingo, the capital of the Chickasaw Nation, that Murray would find enough legal and political opportunity to last him a lifetime.

The land south of Kansas and north of the Red River had been given over to Native Americans removed from their ancestral lands, and, by 1889, more than twenty-five tribes were assigned there. There remained, however, much of the central and western lands used for grazing and hunting, and not designated for any particular tribe. White settlers coveted those lands and, in 1889, Pres. Grover Cleveland signed a bill opening the Unassigned Lands to white settlement. A year later, Pres. Benjamin Harrison signed the Organic Act establishing a government for the Unassigned Lands, which became known as Oklahoma Territory.[13]

In 1898 (when Murray arrived there), Indian Territory and Oklahoma Territory existed side by side, like two rattlesnakes cuddling. Both territories were seeking a pathway to statehood and neither wanted to cede any autonomy to the other.

Unlike Oklahoma Territory, Indian Territory didn't actually belong to the United States and was not a part of the political and legal makeup of that nation. Each tribe (or confederacy) in Indian Territory governed itself, made its own laws, and set the penalties for breaking those laws. Whites continued to pour into both of the Twin Territories, but they were at a severe disadvantage when it came to settling in Indian Territory. Non-citizens could not own land or engage in some professions; their children were barred from tribal schools. Eventually, political pressure on Washington from white settlers led to passage of the Dawes Act, empowering a committee to convince the Five Civilized Tribes to accept individual allotments of property in exchange for giving up the tribal lands. Treaty negotiations went nowhere for six years. Citing lack of progress, Congress then directed the Dawes Commission to begin enrollment for allotments whether the tribes agreed or not. Indian Territory was at risk of losing its autonomy altogether.

When, in March 1898, William Henry Murray stepped off the mail stage in Tishomingo carrying a soiled carpetbag, a copy of the US Constitution, and his last dollar, he stepped into a chuckwagon stew of governments, legal systems, laws, and politics unlike anything ever seen in a century and a half of US history.

He settled in and made himself at home.

Tishomingo was a small but bustling town—fewer than five hundred residents—in rolling black dirt country on Pennington Creek twenty miles or so north of the Red River. The nearest rail connections were at Ardmore, thirty miles to the west, and Durant, thirty-five miles southeast. Tishomingo had a bank, a blacksmith shop, a barber, a hotel, and the usual assortment of small-town retail businesses, all clustered along Main Street and fronted by board sidewalks. Nine months a year the streets threw up clouds of dust; three months a year they were six inches deep in mud.

The most prosperous building in Tishomingo was a massive three-story red granite block structure with cones and arches of the Romanesque style. Located atop a hill overlooking Main Street, the new Chickasaw National Capitol Building was occupied by offices of the governor and both houses of the legislature. Tribal disputes involving civil and criminal matters were tried in its courtrooms, and territorial court met there.[14]

An educated man with a law ticket and small-town manners could find plenty of work in Tishomingo. Within weeks, Murray affiliated with a local law firm and was trying cases. He involved himself in politics as well. The newcomer landed in Tishomingo in the midst of a bitter campaign for the office of governor of the Chickasaw Nation. Murray's law firm was working for Douglas H. Johnston, candidate of the National Party and supporter of private land ownership. Murray signed on to serve as party secretary and, when the party won across the board, earned the friendship of the new governor of the Chickasaw Nation.[15]

Murray's friendship with Governor Johnston won him plenty of legal work and requests to help draft bills for the Chickasaw legislature. Murray recommended a new tax structure that helped free the Chickasaw Nation from a $300,000 tribal debt. Despite the workload, Murray continued to read and study. He had memorized the US Constitution word for word and could recite long passages from Peter Parley's *History of the World.* Some of the first money he earned from his legal work in Tishomingo went to retrieve a crate of books—his personal library—that he had stored in Fort Worth.

As a young man-about-town—a man with some admirable political connections—Murray was expected to join in the parties, dances, and dinners of Tishomingo's social elite. It was at one of those events that Bill Murray met Governor Johnston's niece, Mary Alice Hearrell, and they were married on July 19, 1899.

Mary Alice was twenty-two years old, a teacher at the Bloomfield Academy, and a tribal member. She was an attractive, clever young woman; Murray was a successful attorney with plenty of promise. Typical of the times, their relationship was marked by a Victorian formality—she was "Mother" and he was "Mr. Murray"—and a lifelong comity. Murray kept her removed from his political activities, but she never failed to provide sanctuary for his ambitions.[16]

The marriage cemented Bill Murray's place among the most influential tribal members, and that led to still more tribal legal work. The marriage also gave Murray tribal membership as an "intermarried white," which allowed him to own property in the Chickasaw Nation. For the first time in his life, Murray was well off financially. His legal work for the Chickasaw Nation and Governor Johnston allowed him to earn far more than the typical small-town attorney. In 1901—with one child and one on the way—the couple purchased a large farm east of Tishomingo.

Murray had always had an affinity for the soil; he was born on a farm and supported himself as a youngster by farm labor. Now, with property of his own and leisure time to study, he could become a *scientific* farmer, testing to determine the best crops under existing environmental conditions. As usual, he threw himself into studies of agriculture, soil analysis, climatology, market histories, and transportation networks. In the end, he determined that alfalfa, a clover-like forage crop, was the optimal crop for the land north of the Red River. Also typically, he felt compelled to share his knowledge with others, and began making speeches to farmers about how to grow the lucerne. In reporting one of the speeches, a small-town newspaperman referred to the speaker as "Alfalfa Bill." Soon after, a statewide farm publication asked Murray to contribute some articles, and then published them under the headline "Alfalfa Bill on Alfalfa."[17]

Thereafter, William Henry David Murray would be known as "Alfalfa Bill."

"Alfalfa Bill" Murray in 1905 was far stronger timber than the green stick that had arrived in Tishomingo seven years earlier. At thirty-six years old he was married, settled, and the father of three boys. He owned a farm with tenants and a library of more than five hundred books (which he kept shelved in a log "office" built on his property). His law practice provided a comfortable living, and he continued to be an organizer and a joiner.

Arriving in Tishomingo, Murray gave up the long wool Prince Albert coat and bowler hat popular in Fort Worth in favor of the loose white cotton suits more common in Tishomingo. He was gaunt—stringy as a celery stalk but muscular—5 feet 10 inches tall at 142 pounds, though he looked taller because of his long arms and legs. His face was long and oval, which only accentuated large, droopy ears. Murray had an orator's voice, deep and booming, that could carry to the farthest table of a farmers' lodge hall or the back row of a large municipal auditorium. His eyes were deep set; he cut his hair short in the current fashion and wore a wide handlebar mustache.[18]

Murray had an uncanny knack for cultivating the friendship and loyalty of those he met socially, through his legal practice, in political organizations, or who attended any of his many speeches in the Territory. Friends came to recognize, however, that there were times when Alfalfa Bill would march downhill from the capitol or along the Main Street boardwalk with his eyes cast down, looking neither left nor right, white suit flapping around his lanky body like a flag at half-staff. At these times, he was deep in thought, processing new information, organizing new concepts in his mind.[19]

At the beginning of 1905, Murray was considering the nature of statehood and constitutions.

As non-Natives in Oklahoma Territory organized for statehood, there was considerable resistance to the prospect in Indian Territory. Although most observers expected that the Twin Territories would enter the United States as a single state, those in Indian Territory felt that a two-state entry would preserve some of the autonomy promised to the Five Civilized Tribes time after time. Residents of Indian Territory called for a convention of their own to establish the Indian state of Sequoyah, to be held in Muskogee in August 1905.

Murray was elected representative to the Sequoyah Convention from Tishomingo and had a hand in choosing delegates from the rest of the Chickasaw Nation. Prior to leaving for Muskogee, he collected and studied constitutions from most of the other states and at least three other countries. Murray gave the keynote address, was elected to represent the Chickasaw delegation, and served on the committee to help draft the new constitution. The voters in Indian Territory needed only to ratify the new constitution and present it to Congress for the state of Sequoyah to be born.[20]

The new constitution was ratified on November 7, 1905, and a delegation was sent to Washington to ask for separate statehood. It was there they learned the facts of political life: if the Twin Territories entered as two states, Congress would have to seat four new (likely) Democratic senators, a significant change in the balance of power in that body. The single-state option would mean only two new senators. In June 1906, (Republican) President Roosevelt signed legislation providing for the admission of Indian Territory and Oklahoma Territory as one state.[21]

The Sequoyah convention was no failure, however. It created a cadre of experienced men with statewide political ambitions, strong local support, and knowledge of how governments are created. These men would dominate Oklahoma politics for the next half-century.

Once again, districts in the Twin Territories went to the polls to elect delegates for a statehood constitutional convention; and once again, Alfalfa Bill Murray was elected as the delegate from Tishomingo. Oklahoma's statehood constitutional convention opened in Guthrie on November 20, 1906, and the Sequoyah men had already plotted their strategy. Taking advantage of Murray's support by the Democratic Party and the Farmers' Alliance—and with the Sequoyah men voting as a bloc—Alfalfa Bill was elected president of the convention by an overwhelming majority.[22]

"There are thirty-five old Sequoyah constitutional delegates in this convention," one newspaper groused. "They wrote the Sequoyah constitution and the indications are that they are going to write the Oklahoma constitution."[23]

Murray and the Sequoyah group dominated the convention, furthering their own social philosophies while gathering political allies from across the state. From the start, Murray made clear what he wanted to see in the new state charter: protection for farmers from banks and transportation companies, restrictions on foreclosure, prohibition of land ownership by aliens, agricultural and mechanical training for all ages, and other elements codifying his agrarian philosophy that "civilization begins and ends with the plow." Along the way he worked to assure his own political future. Murray appointed forty-three committee chairmen, each chairmanship a political favor. Alfalfa Bill was also responsible for determining county boundaries and county seats for the new state, providing another opportunity for political carrots and sticks. (This earned Murray another nickname: "Cockleburr Bill," given by those who thought his handling of the convention was too rough.)

Generally, his efforts were appreciated, but he didn't always prevail. A gaggle of Guthrie society women attended the convention and presented Murray with a floral bouquet. He halted the session in progress to give an impromptu speech. "All the bouquet lacks is a few stems and blossoms of alfalfa," he said. "Why not adopt that plant as the floral emblem of Oklahoma? Alfalfa—which sends its roots deep into mother Earth, to water's depth, and shoots its succulent stems upward, leaving the soil richer and serving man as food for his herds. It is the emblem of perpetual institutions and wholesome laws. Let us call this 'The Alfalfa State!'" (The convention chose mistletoe instead.)[24]

Murray had taken the leadership role he had always coveted and forged a constitution for a new state of the United States. The constitution was ratified by a two-to-one margin on September 17, 1907. Two months later, men and women of the Twin Territories gathered at telegraph stations, churches, and public buildings for word from Washington. On November 16, President Roosevelt signed the statehood proclamation and telegraphed the news to those who had been anticipating it.

"Oklahoma is a state," Roosevelt declared. Alfalfa Bill Murray was one of its founding fathers.

The Sequoyah men had enough political clout to claim almost any elective office in the new state. Alfalfa Bill considered standing for the US Senate, but deferred to another Democrat from the former Indian Territory. Instead, he chose to run for a seat in the new state legislature. Murray took his seat in the Oklahoma House of Representatives on December 2, 1907, and was immediately elected speaker of the house. There were plenty of budgetary and housekeeping details to be worked out during that first-ever legislative session. Murray kept the legislators focused and productive, and he wasn't shy about taking the podium for one of his interminable lectures about government. Some complained about Murray's verbosity, but when the governor visited the House prior to adjournment, he reminded the legislators that "from among your membership you selected a Speaker. It is not a misrepresentation, he *is* a speaker."[25]

By the end of the first legislative session, Murray had spent three years in conventions or the statehouse; his law practice was in shambles and he was deeply in debt. When he moved his family back to the farm, there was one additional member: a daughter, Jean. Alfalfa Bill dusted off his law books and set about rebuilding his financial resources. But he could stay out of the political arena for only so long. Within a year, he was already planning a gubernatorial campaign.

Murray's 1910 race for the governor's office never caught fire. By 1910, Oklahomans were flush with the benefits of statehood. Crop prices were high and the small farmers—his source of strength for so many years—were indifferent to his campaign. He ran a close second in a four-man race. The loss didn't discourage him; he was already looking at the national stage.

Given the nation's increased urbanization and industrialism, Murray was more convinced than ever that the preservation of rural America and the sanctity of the family farm were key to the future. In 1912, he announced plans to run for Oklahoma's new at-large seat in the US House of Representatives. The statewide race was clean and straightforward, and Murray won in a walk. The family he would leave behind when he moved to Washington had grown even larger with the birth of a fifth child, a boy, born during the campaign.

Alfalfa Bill was one of 435 members of the House and far from being a leader of anything. There was only so much farming legislation to consider, so he involved himself with immigration, Mexican affairs, international banking, and several other topics, and the time he spent propounding on them exceeded the amount of time spent studying the subjects. The topic of interstate highways and road financing came up often, and Murray's position was consistent. He was in favor of federal aid for road building, but only for farm-to-market and post roads. Federal aid for "peacock boulevards" would be a waste of money, he said, benefiting only the "automobile tourists."[26]

Murray almost lost his 1914 bid for reelection. Alfalfa Bill made a lot of friends on the national stage during his first term in Congress, but he accomplished little. A particularly long session prevented him from returning to Oklahoma to campaign, and farmers were beginning to question whether Bill was doing all he could to represent them. "He received less than one-third of the Democratic votes cast in the district," a newspaper wrote. "We would not call that a very rousing endorsement of Bill's record."[27]

Though he spent just $500 on his reelection campaign, he was, once again, short of cash. The solution was an invitation to join the Chautauqua circuit, a traveling tent-show lecture series that brought culture and entertainment to small towns across the United States. Alfalfa Bill was offered $85 a lecture, and he booked as many as possible. (He spoke fifty times in sixty-four days, earning a quick $4,250.) He was billed as "a homely philosopher with a wonderfully big and resonant voice." He spoke on "The Philosophy of the Plow," sharing his complex theories of nature, history, and humankind with farm families in a droll and homespun manner.[28]

"Being a farmer [Murray] goes in for comfort rather than style," one attendee wrote. "While here he wore a crash suit, a little the worse for wear, and a linen duster that came to his shoetops. The big Oklahoman does not consult the fashion plates before appearing on the platform or in public."[29]

The Chautauqua speeches (and his regular addresses to the National Farmers' Union conventions) gave him a national reputation among farmers. The national press in Washington found in Alfalfa Bill a very smart rural hayseed who was always good for an outlandish quote on a slow news day. A

Washington columnist heard that Bill was dissatisfied with the flower boxes of begonias, geraniums, and pansies planted around the capitol.

"I'm thinking of asking the Secretary of Agriculture to tear out those weeds and sow alfalfa instead," he told the columnist. The congressman mopped his brow with a big red bandana, rubbed his walrus mustache, and readjusted frayed galluses. "The next thing we'll need is a silo, and that's only one step removed from a big red barn."

The columnist filed his story under the headline "Bill Balks at Begonias."[30]

As Congress returned to session in 1914 the European nations were going to war and politicians had to deal with issues of neutrality, preparedness, and intervention. Murray had given the issue much thought; he'd studied the geopolitical, economic, and cultural histories of the nations and concluded that the war would continue until all of Europe was reduced to rubble or until the United States entered the war to tip the balance. Murray's clamorous restatement of this conclusion put him at odds with other Democrats, particularly Woodrow Wilson, who was campaigning for reelection under the slogan "He kept us out of war."

Murray was labeled an interventionist, and his break with Wilson marginalized him for the rest of his term. The farmers of Oklahoma still felt Murray's agrarian philosophy benefited them, but when Murray spent all his time on preparedness issues, the farmers rejected him.[31]

In the fall of 1916 Murray failed to win reelection for his third term in Congress; Woodrow Wilson was reelected for a second term. Six months later, the United States declared war on Germany and entered the Great War.

Friends wouldn't allow Alfalfa Bill to live the life of a private citizen. Bill, on the other hand, was frustrated at the powerlessness of national office and cynical about a fickle electorate that voted its own narrow self-interest. Nevertheless, Murray allowed his supporters to organize a campaign for governor in 1918.

There were no strong issues on which to base a campaign, no reason for voters to make a change in the statehouse. Murray ran a polite and respectable—if indifferent—campaign and lost.

Murray returned to his farm, determined to give up politics forever. The aches and pains of age were creeping up on him, and he began to experiment with different diets, which led him to a detailed study of nutrition and anatomy. He changed his diet to include mounds of unprocessed greens and twenty-five to thirty glasses of buttermilk a day.

"I discovered in my reading on foods that the mental exertion for the four years I was in Congress produced a condition which necessitated tropical sun," Bill said. "I began to think about traveling to the tropics."[32]

His further studies of tropical climes led to a full-blown plan to create new farming colonies in Bolivia. In 1924, Murray cashed out everything he owned in Oklahoma and led eighty-two colonists (including his sons and their families) to their new homes in South America.

Forty-five years old, W. H. "Alfalfa Bill" Murray would be starting over from scratch in Bolivia.

Alfalfa Bill and his wife returned from Bolivia with $45 to their names, his bid to create a new land of yeoman farmers as dead as a heart-shot deer. (He was unable to pay the fare for his three sons to return to the United States. They would stay in South America, selling off farm gear from the failed colony, until they accumulated enough money to return.) Newspapers speculated on when he would arrive in Tishomingo, and his former neighbors were excited about the return of their native son. "If he will stand for it, the band will blare and the people will extend him the glad hand." Hundreds of admirers were waiting at the depot when he arrived on August 24, 1929.[33]

Without a home of their own, Murray and his wife, along with their two youngest children, settled in at his sister-in-law's home, agreeing to pay for rent and groceries. According to one story, Murray set off right away with what cash was left. "To that, he added forty dollars borrowed from the bank, established credit at the grocery store for his family, gave his wife a dollar, and fared forth to run for governor."[34]

Alfalfa Bill wouldn't announce his decision to run for office for four more months, but he toured Oklahoma giving speeches on Bolivia (ending each with his favorite theme, "Civilization begins and ends with the plow"). At each event he accepted a modest honorarium or allowed the sponsor to pass the hat on his behalf. Most of the money he sent back to his family in Tishomingo.

The tour allowed Murray to take the political temperature of the state, and the political landscape of 1929 was quite a bit different from the one he had left five years earlier. Conflicts over Prohibition and the Ku Klux Klan weakened the Democratic Party and allowed the election of the state's first Republican senator. Partisan squabbles in the statehouse diluted the influence of Bill's old Sequoyah allies. The one political leader unsullied after the political brawls of the previous decade was Alfalfa Bill Murray. In October, two months after his return to Tishomingo, the stock market collapsed, bankers retrenched, and farmers—Alfalfa Bill's natural constituency—saw crop prices tumble.

Murray announced his candidacy for governor in January 1930.

What Murray did in the first few months of 1930 could hardly be called a campaign. He was virtually penniless; he again relied on small honorariums or voluntary collections to pay for a cheap hotel room or a diner meal. A few old friends—acting more out of pity than with any hope he could win—took time away from work to drive Alfalfa Bill from speaking engagement to speaking engagement. A dairyman from the eastern part of the state sent Murray a big wheel of cheese. He carried it with him from county to county, always wrapped carefully in one of his red bandanas to protect it. He could usually buy a few crackers to go with the cheese, wash it down with some fresh water, and call it a meal.[35]

Murray spoke to the farmers, avoiding the big cities and hopscotching from small town to small town. He spoke of cutting government operating costs and adjusting the state income tax. "Then I spent much time discussing the principles of gross income tax and taxes in general," he said.[36] Alfalfa Bill didn't talk *to* the electorate, he talked *with* them, explaining how the current taxing system was threatening ownership of their homes and acreage.

If Murray bypassed the big cities, the big-city newspapers ignored Murray. Yet, in every small town, Alfalfa Bill found time to visit the office of the local newspaper and, lanky legs draped over a chair next to the stove, talked with the editor for hours about farming, local crops, roads, and the current financial crisis. Endorsements from small-town newspapers stacked up.

The frontrunner of the seven Democratic primary candidates was an oil millionaire who was grateful for the humor a character like Alfalfa Bill could bring to an otherwise dull race. The oilman played Murray for a rube, with his unpressed hand-me-down trousers, collarless shirts, and variety store socks. The attacks backfired, especially among rural folk who dressed more like Murray than the millionaire.

Even as the rural precincts rallied to support him, Murray's campaign was treated with tolerant amusement by urban politicos. It wasn't until the primary votes were counted—with Alfalfa Bill the top vote getter with 49 percent of the votes—that the big-city newspapers began to take him seriously.

Murray consolidated his support during the run-off campaign, managing to win all but Oklahoma and Muskogee counties. On November 4, 1930—the same day Ross S. "Fat Boy" Sterling was elected governor of Texas—William H. "Alfalfa Bill" Murray was elected governor of Oklahoma. It was a victory celebrated in poetry by a man from Claremore:

> In spite of his lack of the stuff called jack
> To pass among his backers,
> Bill won the race while feeding his face
> On cheese and sody crackers.[37]

Texans and Oklahomans chose two very different men to lead their respective states.

Ross Sterling was a businessman, the first businessman elected to that office in Texas. Voters looked at Sterling's experience in retailing, banking, oil production, and investment, and trusted him to solve the growing problems of business closures, bank failures, and unemployment. Bill Murray was a farmer/philosopher, a whip-smart political figure who stood up for the needs of families trying to scratch a living out of the earth. Sterling knew how to employ capital to solve problems; he knew how to delegate important issues to smart people. Murray approached problems as an attorney does: using logic and precedence to advocate, communicating clearly and confidently, and taking immediate action with any tool at hand.

Sterling was elected for his competence; Murray was elected for his imagination.

As 1930 came to an end and the governors-elect awaited their inaugurations, construction continued on the new free bridge over the Red River. The *Dallas Morning News* noted, "Rapid Progress on Bridge near Denison." The first of five steel sections had been installed several months ahead of schedule. The *Daily Oklahoman* reported delays in construction of the southern approach roads, but the occasional delay was to be expected.[38]

Meanwhile, every farmer, businessperson, family, and tourist continued to pay seventy-five cents to cross the Red River on the old toll bridge, all of them looking forward to the day when they could travel freely.

CHAPTER 6

So Much Legal Gobbledygook

SIX MILLION unemployed Americans greeted the New Year in January 1931.[1]

The nation was in trouble. Somehow—in a manner many Americans still didn't fully understand—a disastrous day on New York's Wall Street fourteen months earlier had resulted in the unemployment, plummeting crop prices, bank failures, and absence of credit that were strangling Texans and Oklahomans.

"The Depression had hit Texas hard," governor-elect Ross Sterling admitted. "Depression clouds had all but blacked out the sun of optimism."[2]

Oklahoma's labor commissioner estimated that there were five thousand unemployed in the state; the reality was probably four times that. A similar survey in Texas was abandoned when men refused to answer the questionnaire because of shame.[3]

Spot cotton opened the year at $9.66, down $6.00 from when it was planted. South Dakota corn was down to three cents a bushel by the end of 1930. Falling crop prices meant farmers could barely afford to harvest the crops they put in the ground; small-town newspapers across the Midwest published the old joke about a farmer who broke his back trying to lift twenty-five cents' worth of wheat. Warehouses were filled to overflowing with products no one could afford to buy.

The price of oil continued to fall as production increased. Farm acreage and oil storage tanks used as collateral for loans were worthless, and more banks failed. (Thirty-six hundred would fail before summer.) Every bank failure led to more unemployment as credit dried up or business-operating cash disappeared. When the bank at Picher, Oklahoma, failed, the local school district lost all payroll and textbook money, forcing the district to fire its teachers and cancel the school year.[4]

Entire communities tottered on the knife-edge of panic. The food riots began on January 3 in England, Arkansas, when five hundred desperate farmers congregated in the business district, threatening to take food by force from merchants there.[5]

"Our children are crying for food and we're going to get it," a man in the crowd shouted. "We are not going to let our children starve," shouted another through tears, even as he waved a pickaxe. "We want food and we want it now."

Farm families in the once-prosperous community had been ravaged by a yearlong drought, bottomed-out cotton prices, and failure of the Citizens Bank and Trust a month earlier. "We're not beggars," another in the crowd stepped forward to explain. "Give us work and we'll not come back."[6]

Town merchants wired Red Cross headquarters in Little Rock, which promised to reimburse $2.75 per family to pay for provisions. Well into the night, rioters stood in orderly lines at England's stores, waiting to "shop" for their share of donated goods.

The United States in 1931 had no unemployment insurance, no Social Security; county poorhouses or the open road were the only options for many. The economic problems were overwhelming and unrelenting, and people looked to their governments for answers.

Pres. Herbert Hoover was, like Ross Sterling, a self-made millionaire. He was elected in 1928 on a prosperity platform, and he was hesitant to acknowledge the depth of the economic crisis in 1931. "Ninety percent of our difficulty in depressions is caused by fear," he said. "There should be confidence, for our country is fundamentally sound."[7]

Drawing on his extensive business experience, Hoover believed the mainspring of the relief and recovery effort was a generous citizenry. Failing there, the effort might call on existing charitable organizations, such as local churches, the Red Cross, or the Salvation Army. Meanwhile, he asked Congress to provide loans to banks, railroads, steel companies, and other too-big-to-fail business segments.[8]

Drastic conditions soon overwhelmed private charities. In the wake of the food riot in England, Red Cross officials in Arkansas surveyed the rest of the state. About 100,000 persons were currently receiving assistance in that state alone, and they predicted the number might increase to 250,000 by the following winter. The American Red Cross chapter in

Dallas responded to a request for help from drought-stricken Clarksville by shipping twelve crates—each weighing three hundred pounds—full of canned foods and used clothing. This one shipment exhausted the chapter's supply of used clothing. A church group in Texas closed its nickel soup kitchen in less than a week, recognizing it could never meet the high demand with only donated food and volunteer labor.[9]

Governor Sterling's inauguration speech in Austin on January 20 made little mention of the financial hardships Texans were facing. Like President Hoover, Governor Sterling felt that the "major cause (if not the primary cause) of the whole trouble was a collapse of public confidence." He was averse to any form of the dole, suggesting that "those who don't work shouldn't eat." He endorsed the Texas Press Association's "Give a Neighbor a Job" ad campaign, in which families were encouraged to hire out-of-work neighbors to do odd jobs around the house. The sober, steady, business-leader-now-governor summed up his plan to deal with economic hardship several days after the inauguration: "Don't worry about the hard times," he later told a Fort Worth audience. "Just go to work."[10]

Governor Murray's extemporaneous inauguration speech was a laundry list of specific actions that the new governor would take to ease the plight of Oklahoma farmers. Standing on a makeshift platform decorated with a few flags, he pledged aid for "the farmer at the plow without seed to plant"; "the condition of labor and its hungry children"; and "those hopelessly tramping the highways." Within days, Murray went before the state legislature, demanding "an appropriation for the purchase of free seed—field and garden—for the tenant farmers who are unable to purchase seed for themselves, and who are without credit." He said he would suspend the law requiring counties to do roadwork only under contract to companies, "so that local labor could be employed by the day." Further, he wanted a reduction in the state's ad valorem taxes and in state budgets. "For years the cost of government has been more than it's worth," he snorted.[11]

Eight days after Alfalfa Bill's inauguration, food riots broke out in Oklahoma City. Nearly a thousand unemployed men and women, shouting that they were hungry and jobless, marched on city hall, looting several stores on their way. Police dispersed the crowd with fire hoses and tear gas, but not before twenty-six were arrested.[12]

Food riots and hunger marches continued throughout the nation that winter. A group of sixty marchers in Los Angeles were gassed and beaten before being dispersed. Two thousand homeless men in San Francisco were in line at a soup kitchen when police attempted to arrest several speechmakers in the crowd. A five-hour street brawl ended in the death of one man and the arrest of scores of others. An anonymous letter writer threatened to dynamite El Paso city hall unless work was provided to the city's two thousand jobless. Armed mobs marched in Amarillo, Duncan, Austin, and elsewhere in Texas and Oklahoma.[13]

Meanwhile, local banks kept toppling like dominoes. The American National Bank in Paris locked its doors on the afternoon of February 4 and never reopened. Athens National Bank posted a notice on its locked doors, saying the bank would be involuntarily liquidated. Kindly Mr. McCrory of the bank in Ringling, Oklahoma, locked up one afternoon and left town before supper for an extended vacation, never to be heard from again. The banks in May, Beggs, and Enid, Oklahoma, failed within days of each other; Abilene, Archer City, and Austin, Texas, banks were shuttered without warning, leaving thousands of customers with nothing.[14]

Americans everywhere were hurting, and it seemed there was little their governments could do (or would do) to help them.

Conditions in Durant, Denison, and communities on both sides of the Red River near the toll bridge were no worse (but certainly no better) than in other sections of the country.[15]

In 1931, an evening drive through the residential streets of those towns (and others like them) might disclose which families still had jobs and which were unemployed. If the light behind the curtains showed white and steady, chances were that the breadwinner still had a job and could pay his dollar a month to the electric company. But if the light was yellow and flickery, that was a house where the power was off and the father had pulled the old kerosene lamps out of his shed.

In the home of the employed, life might have been much as usual, with birthday parties, church suppers, and an occasional trip to the Rialto for a double feature. But inside, the home would be painted with dread. The husband would put off buying a new work shirt, knowing his wife could replace the frayed cuffs. And each day the wife would set aside string and wrapping paper from the butcher's bundles, storing it in an enameled kitchen cabinet, knowing it might be reused. Each night they would go to bed, praying his job as shopkeeper, factory worker, or clerk would still be there at the end of the month when the mortgage payment came due.

Shame lay like a hot stone on the home of the unemployed, and fear radiated outward in waves. Nothing was as usual. There was no electric refrigerator, no money to buy ice, and no money to buy meat and eggs even if there was a way to store them. The husband might spend his days walking around town, looking for work or congregating with other men who spent their days walking. The wife would stay at home, worrying whether the house would go back to the bank and what would happen then. When he came home, the husband would probably say little, carrying a guilt that he couldn't care for his family. She would say little, afraid his shame would turn into anger.

On the near north bank of the river—close to the toll bridge and just past the trusses being assembled for the new free bridge—lived mostly farmers. They grew corn, cotton, and peanuts; and every farm had poultry, a little livestock, and a subsistence garden with—depending on the season—tomatoes, melons, legumes, and maybe spuds. Most of the black farmers were tenants; most of the white farmers were buying their land with every hard-won payment to the bank. Black or white, the farmers were thin, eyes deep set and watery, weathered skin like leavings from the cobbler's floor. Their blue denim overalls were bleached white in places from wear and sun. A farmer is never unemployed, but many of them owed more than they could pay and they wondered where they'd find the money for seed and six more months of groceries.

Businesses in downtown Denison, Durant, and Sherman clung to solvency like a lizard to a garden wall. Dollars seemed harder to come by. Folks hung on to cash, and it was rare for retailers to hear the sound of half-dollars whining and singing as they skimmed across marble sales counters. The barber might have had three chairs in his shop, and two once were busy every day. After a time, there was just one barber, alone in his shop, with the mirrored shelves reflecting colorful bottles of hair tonics, all waiting for men who had decided they could get by a little longer between haircuts. In most stores, selection would be thin and inventories low, another cost of expensive credit. None of the big local banks had defaulted, but the *Denison Herald* carried stories of bank runs and failures as close as Sapulpa, Dalhart, Gainesville, and Poteau, and bankers weren't held in the high regard they once were.

Fewer tourists passed through the Gate City. Out on the state highway—almost any state highway in Oklahoma or Texas—the cook and waitress at the chicken dinner restaurant would have been let go, leaving the owner and his wife to handle the big skillets and the tables. A live-over gas station would be making most of its money selling cold drinks and salted peanuts to road and bridge crews working nearby; the mechanic sat idle.

On a day in March 1931, four thousand vehicles crossed the toll bridge. Twelve hundred of them were vehicles with out-of-state plates or with Texas or Oklahoma plates from out of the area. The twenty-two hundred remaining trips across the toll bridge—north to south and south to north—were taken by a woman delivering a wedding gift to her niece across the river, a farmer making a loan payment at his bank, a family driving to a rare outing at the picture show, a couple shopping for a bassinet, a banker looking over collateral for its loan worthiness, people going to a job, or people looking for one. To drive the thousand feet across the toll bridge, every vehicle must stop and every driver must hand over to the toll taker seven dimes and a nickel—the daily toll paid by weary and frightened people who worried about having two dimes to rub together.[16]

However, as winter turned to spring 1931, Denison, Durant, and the communities on both sides of the toll bridge over the Red River had a reason for hope: completion of the new free bridge.

"More puzzling than vexatious is the evident purpose of Texas authorities to keep the Red River toll bridges in operation as long as possible," the *Daily Oklahoman* wrote. "Why the officials down Austin way should prefer another summer of toll bridge exactions to the free bridge accommodations across the boundary river is beyond understanding."[17]

By April 1931, Oklahoma Highway Department officials were smelling a rat; Texas was dragging its heels in completing work on the free bridge north of Denison.[18]

The bridge was all but complete by March 1, lacking only silver paint on the steelwork and a sign-off by state bridge inspectors. Texas, though, had not yet completed the highway from Denison to the bridge's southern approach road. "It is believed that the road from this city to the bridge will be ordered paved by the State Highway Commission during its April meeting," according the *Denison Herald*. Eighteen months after the states agreed to build the bridge, Texas hadn't even let the contracts to build the roads.[19]

"Is Texas merely trying to keep her people from entering Oklahoma and becoming subject to the superior attractions of a younger and more attractive state?" Oklahoma newspapers jeered. "It is the riddle of the hour."[20]

The "riddle of the hour" was no riddle at all to commissioners of the Texas Highway Department. A year earlier, as chairman of the Texas Highway Commission, gubernatorial candidate Ross Sterling signed a contract with the Red River Bridge Company meant to end litigation, allow construction of the new free bridge to proceed, and (not coincidentally) take the issue off the table during Sterling's campaign. Sterling and the Texas Highway Commission promised to pay the bridge company for each month the toll bridge lost when the new bridge was completed and a cash payment when the free bridge opened. Texas also promised that the new bridge would not open before July 1.

It was a contract never meant to see the light of day.

A new member of the Texas Highway Commission (who likely knew nothing of the contract) responded to the complaints of delay, claiming that Oklahoma rushed work on the structure to embarrass the Texas construction company. He implied that Oklahoma was willing to open the bridge without going through a formal inspection, a step Texas wouldn't allow. Commissioner E. R. Martin said he hoped to see the bridge opened by July 1 at the earliest.[21]

His comments spawned a storm of protest in the Denison area and with the Oklahoma highway commissioners. Denison businesspeople and residents had been waiting three years for their bridge and were frustrated at the slow pace of construction; Oklahomans took umbrage at being blamed for what was, obviously, Texas' dillydallying. Texas highway commissioner W. R. Ely was forced to step up with some clarification. In doing so, he let the cat out of the bag.

"We felt we would not be able to construct or open a free bridge across the Red River for five years," Ely explained, citing the pending and threatened lawsuits that delayed construction. He admitted that there had been an agreement between Texas and the toll bridge owners, and that one of the conditions was that the free bridge would not open until July 1. "We are going to comply with the provisions of the contract and will abide by it religiously," Ely promised.[22]

Never one to pass up an opportunity to rebuke Oklahomans, the *Dallas Morning News* reminded readers "that the long reign of toll collection is partly due to a franchise granted by the Choctaw [*sic*] Nation in which Texas had no share." The newspaper went on to lecture, "A little more patience is justified where the pledged word is to be made good."[23]

Commissioner Ely was not asked about (nor did he offer to disclose) any other provisions of the agreement with the toll bridge company.

For the next sixty days Texas highway officials slow-stepped completion of the southern approach roads. A request for bids to complete four miles of concrete highway from Denison to the Red River appeared in mid-April; an $80,269 contract was let to a Dallas construction company on May 1. (The Texas highway commissioners crowed that the bid was much lower than their engineers' estimates. Texas had recently enacted a thirty-cents-an-hour minimum wage, but the law had not yet gone into effect.)

Roadwork began in mid-May, and the construction crew boss assured a young *Denison Herald* reporter that the paving would be complete and the concrete cured in time for the highway to open to the bridge by July 1.[24]

A firm date was exactly what Denison Mayor Clarence Scott was waiting for. He wasted no time rounding up a committee to plan a bridge-opening celebration that would involve ten thousand people or more from the entire region. He wanted an event that would enthuse people, energize them, and lift the blanket of dread that was smothering his community.

The fourteen-person committee—consisting of chamber of commerce officials, bankers, business owners, and retailers from Denison, Durant, and Sherman—eventually settled on Saturday, July 4, for this long-delayed extravaganza. The holiday weekend would include auto caravans leaving Durant and Sherman and meeting at the center of the new free bridge for a christening with a bottle of Red River water. The celebration would continue at Forest Park in Denison with speeches, a picnic, terrapin races, a watermelon park, fireworks, and a pavement dance.[25]

"The last great barrier between the neighborly relationship of two great commonwealths has been eliminated!" Newspapers wasted no time spreading the jam. "No longer will the citizen of the Sooner State be required to dig deep into his pocket before he will be allowed to pay a friendly call to his neighbor on the Texas side. No longer will it be necessary for the friendly Texan to halt himself and flivver when he reaches the shores of his beloved and historic old river."[26]

By June 1, invitations had been sent to Governor Murray and Governor Sterling, the district's congressmen, heads of the state highway departments and highway commissioners from both states, and selected members of the Texas and Oklahoma state legislatures. Among those invited to speak was newly elected state senator J. J. "Jake" Loy. As a state representative, Loy had introduced and championed the Texas Free Bridge Law, the legislation that led directly to construction of the new free bridge. Loy had been elected to the state senate in November, and his district now included the Texas counties along the Red River.

Loy wouldn't wait until July 4 to have his say about the almost-ready-to-open bridge.

Jake Loy never really said who passed him a copy of the contract, but he certainly had the goods.[27]

Normally, Loy was an easygoing, small-town guy with the Hollywood good looks of actor Pat O'Brien. On June 11, as he entered the office of Attorney General James V. Allred of Texas, he did so with a dark glare.

"I am anxious to know why all this was necessary," Loy demanded of the attorney general as he slapped a sheaf of papers down on Allred's desk. "All this" was the contract between the Red River Bridge Company and the Texas Highway Commission, signed by now-governor Ross Sterling, along with some related correspondence.[28]

"The Highway Department of Texas agrees to pay the Red River Bridge Company $60,000 in cash before the new bridges and approaches are opened," Loy told the attorney general.[29] Further, he calculated that the state would have to pay at least an additional $20,000, based on the monthly operating penalty. Worse yet, Loy had correspondence to show that the amount would be paid out of auto registration fees, a fund that was held sacrosanct to assist counties with road maintenance.

Allred said he was unfamiliar with the agreement (and a subsequent search of the office's voluminous bridge-related files failed to turn up a copy), but expressed doubt that the earlier attorney general's office would have approved such a contract.

That's when Loy handed over the copy of a letter from former assistant attorney general Rice Tilley to Ross Sterling and the other two highway commissioners.[30]

"I have before me contracts between your department and the Red River Bridge Company," Tilley wrote on August 9, 1930, shortly before formal signatures were affixed. "Although I doubt the authority of the department to make these contracts, some statement of the differences between the parties should be made, and we think we should resolve the doubt in favor of your authority."

Tilley's words express skepticism as to whether the commission was empowered to enter into such a contract. He also questioned the legality of the agreement itself. "There are some recitations to the contract not supported by the law, exactly," he concluded, "but these are immaterial. We have, therefore, approved the contract." Tilley's letter seemed to signal that someone had heavy-handed him into approving the contract. [31]

Loy next called on Governor Sterling.

Ross S. Sterling was not a career politician; he was a businessman who had succeeded by working hard and moving quickly, by hitting four licks where the other fellow hit two, by making deals, closing them, handing the papers off to an assistant, then moving on to the next deal. He was an honest man, eager to close deals and settle matters. He was willing to take a shortcut or two if the final result was honorable.

Sterling acknowledged the contracts were made when he was highway commission chairman and said he was "dissatisfied with them as poor contracts, but done in order to make peace." He reminded Loy that the bridge company "could have indefinitely delayed building the new bridges by federal injunctions."

The agreement was nothing but a "treaty of peace," Sterling said, made merely to placate the toll bridge company's concerns and at the same time assure construction of the toll-free bridge.

"There is nothing to be done now regarding the contracts," Sterling continued, "but when the new free bridge is put in operation it will then be up to [the bridge company] to call for a settlement and state its claims. Counsel for both sides can get together and determine what is due the company from the state."[32]

Loy left the governor's office determined to get to the bottom of things.

Dismissive as he may have been about his agreement with the Red River Bridge Company, Sterling knew there was a problem. State revenues were falling faster than spending could be curtailed. (Just that week the governor had begged the legislature to cut $2.1 million in spending.) Though he didn't reveal it to Loy, Sterling knew that the contracts with the Denison, Gainesville, and Ringgold bridge owners—all with similar terms and provisions—could cost the state up to $350,000. Add the cost of building the bridges, and Sterling—the clean-as-a-whistle businessman elected to his first term because of his competence—might have to account to Texas voters for a million dollars or more. One way out of trouble would be to get Oklahoma, Texas' co-beneficiary in the bridge project, to share the expense.

"Opinion is, in some quarters, when it comes time to open the new free bridge," wrote the gossipy editor of the *Durant Weekly News*, "the Texas Highway Commission will ask Oklahoma to split on the loss. And it comes from other quarters that, when such is asked, a blank refusal will come from the Oklahoma commission."[33]

Before leaving the capitol, state senator Jake Loy asked the attorney general to launch a formal investigation.

Governor Sterling's words startled Joe Bailey Jr., but they didn't surprise him.

In the year since Bailey had negotiated the contract on behalf of the Red River Bridge Company, he and his client had watched the calendar and construction on the new bridge progress. The contract was very precise in saying that Texas would pay the bridge company "*before* the new bridge and the approaches and the roads leading thereto are completed and opened for traffic." With the roads and bridge just weeks away from opening, Bailey's client was expecting to be paid an amount they calculated at closer to $140,000 than the $80,000 Loy estimated for the Denison bridge.

So when Governor Sterling said that *only when the free bridge was completed* could the toll bridge company "call for a settlement and state its claims," Bailey realized that someone might be about to pull a switch on his client. The attorney wasn't all that surprised, however. For several months he had been hearing rumors from friends in Oklahoma that Texas was having second thoughts about the costs of the bridge deal. Once the free bridge opened, the value of the Denison toll bridge would be nil. All negotiating power would be lost.

Bailey couldn't allow the free bridge to open for business until Texas paid up.

On Wednesday, June 24, Little Joe took the train to Houston, to the federal courthouse where Judge T. M. Kennerly of the Eastern District of Texas was holding forth. Bailey had his paperwork prepared and he presented it to the judge. That evening, before leaving Houston, Bailey sent a telegram to Austin.

At ten o'clock the next morning, nine days before the free bridge was scheduled to open, the secretary of the Texas Highway Commission signed for receipt of Bailey's telegram: "Please advise members of the commission that Judge Kennerly, sitting for judge of the Eastern District of Texas today granted an application for restraining order forbidding the opening of the free public bridge across Red River north of Denison on Highway 75."[34]

Word spread quickly. A federal court was prohibiting the Texas governor, highway commissioners, and highway department from opening the new bridge until a hearing could be held on July 3.

Joe Bailey Jr.'s meeting with Judge Kennerly in the Houston courtroom was an ex parte proceeding. That is, he didn't provide notice to the other side to appear. It's a process most often reserved for urgent matters in which requiring notice would subject one party to irreparable harm. When Bailey returned to his Dallas office he wrote a personal and confidential letter to Attorney General Allred explaining why he chose that method.

"I would have made no application for the ex parte injunction without giving you notice," Bailey wrote, "except that I understood very definitely that the Highway Commission of Texas wanted this done in order to use it as a means of compelling the Oklahoma Highway Commission to contribute part of the money necessary to open these bridges." The commission couldn't be open about the reason, Bailey continued, "but the request came to me in such a way that I felt it to be authentic."[35]

Within three days, Texas Highway Department workers placed sawhorses across both ends of the free bridge. Some of the southern approach road concrete was still green and the sand placed over it was not yet ready to remove. The barriers were temporary, but so was the restraining order.

A consensus among public officials and news writers was that the injunction wouldn't hold for long. "The bridges have involved expensive construction. Neither Oklahoma nor Texas anticipates that they will stand idle."[36]

At a meeting on June 29, Denison mayor Scott's committee postponed the opening celebration until July 15, allowing time to plan more activities. The event had outgrown Forest Park near downtown Denison, so they moved it to Munson Park (which, ironically, was named for the brother of one of the original partners of the Red River Bridge Company). The committee arranged for a group of Texas cowboys and Oklahoma Indians to stage a mock battle. State highway patrolmen and more than one thousand Boy Scouts stood ready to direct traffic.[37]

On July 3, Attorney General Allred came before Judge Kennerly to argue that the restraining order should be lifted. "The bridge owners had no ground to stand on in their efforts to prevent opening of the bridge at Denison," he said. Allred attacked the validity of the contract itself, as well as the inclusion of the highway commissioners (including Governor Sterling) as individuals instead of agents of the state.[38]

The arguments put forth were just so much legal gobbledygook to the Red River residents who read them in their newspapers the next day. What was important was Allred's confidence. "He seemed confident of success for the Texas Highway Department in its efforts to thwart the injunction move," according to the *Denison Herald.* Judge Kennerly said he would take the matter under advisement and render his decision on July 9.[39]

The *Denison Herald* assured its readers there were only two possible outcomes: the judge would dissolve the injunction or the state would pay off the contract price to open the quarter-million-dollar bridge. The *Sherman Democrat* agreed: the courts and the state of Texas would protect the needs of the people, people who "see no reason why their property rights in the free bridge, as taxpayers, should be damaged by further delay in use."[40]

In preparation for the July 15 opening celebration, local businesses decorated the fronts of their buildings with streamers, crepe paper, and bunting. Work crews scraped the last of the sand off the approach roads. Newspapers prepared special editions, and the mayor's celebration committee upped its budget for fireworks, hiring a Fort Worth company to build fire pictures of Governor Sterling, Governor Murray, and the crossed flags of Texas and Oklahoma. The finale would consist of a thirty-foot American flag rendered in fireworks and twenty-five mortars erupting simultaneously along the length of the new bridge. According to one committee member, the event would be "the biggest celebration ever pulled off in this part of the country."[41]

Judge Kennerly's decision, delivered Friday morning, July 10, was direct and to the point. "I cannot allow the respondents, the members of the highway department of Texas, to destroy and strike down [the value of the toll bridge and the interests of the toll bridge owners] by opening the free bridge until they pay the sums of money agreed to be paid under the terms of the contracts." The judge found that all or any of the parties were likely to open the free bridge without compensating the toll bridge company.[42]

"When a state officer acts without legal authority," Kennerly wrote, perhaps referring to Governor Sterling, "he is not acting in the interest of the state, and suit against him is not a suit against the state."

In the end, the judge said that the governor, the commissioners, and the state are "enjoined and restrained from in any manner opening or attempting to open for travel by the public the free bridge located north of Denison." Judge Kennerly warned that allowing (or not preventing) travel over the bridge by the public would result in serious consequences to the individuals involved.

Federal judge T. M. Kennerly of the Eastern District of Texas said his action resolved the manner, as far as he was concerned.

The next morning highway department employees replaced the sawhorses with sturdy wooden barriers at both ends of the unused bridge and hammered together a shack for the guard who would be posted there. They found a sign painter to produce a sign on a piece of four-by-six plywood: "Warning! This bridge closed by order of United States District Judge of the Eastern District of Texas on the suit of the Red River Bridge Co. of Texas and its Receiver."

The Denison committee quietly canceled their July 15 bridge opening celebration, "the biggest celebration ever pulled off in this part of the country." There would be no ten thousand visitors. No auto caravan. No bottle of Red River water for christening. No picnic, pavement dance, terrapins, or watermelon park. No fireworks. No fire pictures of Governor Sterling, Governor Murray, or the crossed flags of Texas and Oklahoma.[43]

CHAPTER 7

A Layer of Anger

NEWS OF THE barricaded bridge (and a lack of action by Governor Sterling) aggravated Oklahoma governor Murray. His attorney general and the chairman of his highway commission kept him informed of developments, but Alfalfa Bill had yet to lay hands on the issue. His inaugural address six months earlier was a laundry list of actions to assist stricken Oklahomans, and the governor had been busy working his way through that list.

As promised, Murray distributed free field and garden seed to farmers and cut more than a million dollars from the biennial budget. He promised counties that he would match funds to provide local public construction jobs to the unemployed. Alfalfa Bill browbeat any interest or industry he thought was contributing to economic woes.

"I propose a new 'declaration of independence' from the monopoly of exorbitant prices for bread," Murray announced. The state's bakers were raising bread prices while driving down the price of Oklahoma's wheat. Declaring that the higher cost of bread falls most severely on the poor, Murray threatened to equip the state's prisons with commercial ovens and require that families on relief buy only state-produced bread, priced at only five cents a loaf. The governor had done his homework; after reviewing Alfalfa Bill's inmate business plan, members of the state's bread bakers' association began producing a five-cent loaf.[1]

He faced off with the private utility companies—water, electric, telephone—threatening to take them over and run them with state employees if they didn't lower rates and find some way to accommodate families who were out of work. (A national newspaper columnist wrote about the war with the phone companies for lower rates, joking that Murray had "plenty of support along party lines.")[2]

In July 1931, Alfalfa Bill firmed up support for his program to build five thousand miles of farm-to-market roads, putting local labor to work and purchasing only Oklahoma materials. Enough money had been spent on "tourist roads," according to Murray. Good but inexpensive secondary farm-to-market roads, made of local materials, to connect with main highways point the way to greater prosperity and progress for the state. The labor and production of Oklahoma's farm families will carry us past these hard times, he promised.[3]

Murray was also attempting to corral Oklahoma's oil producers. Oil prices were down to less than ten cents a barrel, and the state's natural resource was gushing away for mere pennies in tax revenue. The governor was unhappy with disappearing tax revenues; producers weren't pleased with the low prices they were receiving. Alfalfa Bill assembled the fractious drillers and forced them to agree to a temporary production shutdown. But the plan would work only if Texas oil producers went along. Surely Ross Sterling, an oilman-turned-governor, would want to join in. Murray presented the plan to Governor Sterling, and Sterling went nowhere with it. For six months the Texas governor consulted his legislature, consulted Texas oil producers, consulted lawmakers again, and generally dragged his feet before telling Murray that Texas wouldn't be joining in. Meanwhile, the mineral wealth of both states was being depleted at an alarming rate.[4]

Sterling was also dragging his feet when it came to fixing the toll bridge mess, and that was cutting Oklahoma farmers out of some lucrative markets.

The Kraft-Phenix Cheese Corporation opened its primary southwestern production plant in Denison just as the free bridge was to open. The plant would produce packaged cheeses, mayonnaise, and dried eggs in massive quantities for customers over a twelve-state area, stimulating egg and milk production by the farmers of northern Texas and southern Oklahoma. Oklahoma producers, however, were at a price disadvantage when they had to pay bridge tolls for every dairy wagon or egg truck traveling across the Red River to the Kraft plant in Denison.[5]

Oklahoma peanuts, a major cash crop for the sandy soil farmers just north of the Red River, were more expensive when delivered to the new Barnhart Mercantile roasting plant in Denison. Imitating the success of the apple growers' association, Barnhart was packaging peanuts and prospecting for unemployed men to sell them for a nickel a bag in the big cities. (On some days, it seemed, the only employment ads in the Dallas, Oklahoma

City, Tulsa, Kansas City, Houston, and New Orleans newspapers were for bill collectors, repossession agents, and peanut vendors.) Oklahoma growers were again at a price disadvantage because of the toll required to get their product across the river.[6]

For Governor Murray, Governor Sterling's inaction served to crank up the heat another notch under a pot that was already reaching the boiling point.

Gov. Ross S. Sterling just couldn't seem to make headway against the economic conditions that were strangling his state. Like President Hoover, he pointed at loss of confidence as the cause, but the Depression was perversely refusing to respond to his hopeful predictions.

The governor tried and failed to get a $350 million multiyear road construction bill through the state legislature. Lawmakers resisted, arguing that the amount was too great and was spread out over too many years to provide immediate assistance. Sterling—no stranger to long-term capital investment—dug in his heels and refused to budge.

His ideas were too often halfhearted and half baked. "I caused some controversy when I suggested employed wives of employed men should give up their jobs in favor of jobless men," he said. Sterling's plan to reduce the number of meals at the state's charitable institutions to two a day was viewed as cruel and out of touch.[7]

Sterling called a conference of business and industrial leaders from all over the state to produce some definitive plans for reducing unemployment. He opened the Austin conference by saying that things might not be as bad as they seemed. The present situation is an extraordinary one, he explained, "in which the people have more of everything, more food, and more supplies, but are faced with more depression. Restoration of confidence in the government and business is vital in dispelling the depression."[8]

His remarks didn't inspire much confidence (or action) among attendees. The conference adjourned without any definitive action or plans. "I believe that many people mistook my reserve for coldness," he later admitted.[9]

Sterling himself wasn't immune from financial collapse. Even as he addressed the business conference he "was undergoing the harrowing last stages of my own transformation from a multimillionaire into a poor man." The Depression caught him, as it did many others, overextended. "I lacked liquid assets with which to meet my obligations."[10]

His personal financial collapse was unknown to all but a few close associates, but the confident businessman who had been elected governor no longer seemed the man to drive a stake through the heart of his state's economic woes.

On top of it all, there was the unfinished matter of the barricaded bridge across the Red River. Every day, his secretary brought him letters and telegrams from residents and booster groups asking him to pay the bridge company and open the new bridge. Editorials in the Denison, Durant, and Sherman newspapers urged Sterling to settle the matter. The *Dallas Morning News* scolded the governor for not paying the toll bridge company. "The toll bridges must go," read the editorial, "but not through expropriation by the state."[11]

Ross Sterling would certainly have been familiar with the Uncle Remus stories from his childhood. Too late, he recognized he was stuck—hand, foot, and head—with a tar baby of his own making.[12]

As governor, Sterling had no approval to release the funds called for in the contract (even if the parties could agree on a negotiated settlement). Jake Loy's Texas Free Bridge Law authorized payment for only *half* the cost of the bridge. Without Oklahoma's participation, The Red River Bridge Company would not receive full payment, and that was likely to cause another round of courtroom delay. In turn, Oklahoma announced plans to take the matter to the US Supreme Court when it resumed hearings in October.

Texas Attorney General Allred was pursuing a new legal strategy to get the bridge injunction lifted (or, at least, move the suit out of Judge Kennerly's court). Allred would argue before the US Circuit Court of Appeals that the contract with the Red River Bridge Company was invalid, entered into without the authority of the state. If that argument prevailed, Sterling knew, he might be held personally liable for terms of the agreement, a debt he could no longer afford.[13]

Worse yet, there had been rumors of violence aimed at one or the other of the bridges north of Denison. Vague and unsubstantiated as these rumors were, Sterling had to take them seriously. The year before, a free public bridge built by Arkansas over the Red River near Texarkana was destroyed by a dynamite bomb the day before it was scheduled to open. State police suspected the owner of a nearby ferry, but no connection was ever proved. On July 8, two weeks after the barricades had been in place on the Denison bridge, the night keeper of the toll bridge north of Gainesville was found shot to death.

No money had been taken, and police were unsure whether the death was related to the man's employment. Sterling passed the rumors to the owners of the Denison toll bridge and ordered a highway department engineer to reinforce security at the new free bridge. The engineer installed two huge acetylene floodlights to illuminate the supporting piers and understructure of the bridge, and Sterling asked the county sheriff to hire local men to stand a twenty-four-hour watch.[14]

Few reporters of the time grasped the complexity of Texas' legal snares and other entanglements and what they meant to the opening of the new free bridge over the Red River. But Governor Sterling knew. He was holding a tar baby that grew stickier and dirtier every time he tried to handle it.

Ask Claud Easterly why he went into the newspaper business and he probably couldn't answer. Son of a railroad carpenter, Easterly grew up in a working-class neighborhood in Denison. He attended Denison High School where he interviewed students, wrote stories, took photos, set type, and edited the school newspaper, the *Denison Hi-Nuzz.* Easterly's father had lost his job in the railroad strike, so college was out of the question for the boy. With his high school diploma and clips from the *Hi-Nuzz,* he went to work for the *Denison Herald* in 1925 as an apprentice reporter for $12 a week.[15]

Still in his teens when he went to work for the paper, Easterly looked older than his years. He had the stocky, compact build of a pipefitter and a hairline that stopped high on his forehead. Easterly wrote up the usual small-town happenings—school board meetings, club gatherings, deaths, church dinners, engagement parties—while earning a reputation for doggedness. He haunted the railroad depot at all hours, befriending ticket agents and telegraphers for news of what notables might be on one of the passenger trains stopping there. At least once he talked a jailer into letting him spend the night in a cell so he could interview a jailed murderer.

Easterly grew up in Denison, attended church there, shopped there, did his job there, and he became as familiar with the ordinary people on both sides of the Red River as a farmer becomes familiar with his livestock. He knew their mood. In July 1931—with that big silver bridge ready for free use but blocked on both ends, and the toll keeper still collecting his seventy-five cents for every passage—the mood was beginning to turn sour.

The young reporter had seen how a crowd's sour mood could turn violent, and the memory haunted him. Fourteen months earlier, he stood on the courthouse lawn in Sherman—the same courthouse where he watched Joseph W. Bailey Sr. argue his final case—as two hundred sullen men milled about, waiting for the verdict in a trial being held inside. A black farmhand had been arrested for a particularly heinous crime against a white woman, and the crowd wanted to be sure lynch justice was done. These were angry men, many of them unemployed, others frightened, and the crowd's temper grew as the day got warmer. By early afternoon the crowd had grown to five hundred or more. A few threw rocks at the courthouse windows. Someone in the crowd hurled a flaming can of gasoline through a broken window downstairs. The judge adjourned the trial and hurried everyone out of the building while peace officers hid the defendant by locking him in the county clerks' fireproof vault as the fire began to spread. The fire department poured water on the burning courthouse until members of the mob used axes to chop their fire hoses into pieces. The fire eventually burned itself out but, even as the courthouse still smoldered, the crowd dug through debris looking for the defendant. The vault was found, forced open, and by sundown the roasted body of the black man was being dragged behind an automobile through the streets of Sherman.[16]

Easterly found the events of that day horrific, made more so by the normality, the familiarity, of the men in the crowd. Already frustrated and frightened by economic conditions too big for a person to solve, these ordinary people turned to anger, lost their senses, became a mob, and committed a monstrous act. Sherman wasn't an isolated case. A month later, a mob in Chickasha, in central Oklahoma, ignored National Guard tear gas to beat down the door of the jail cell holding a young black man charged with assault, and then stabbed him to death. About the same time, a self-appointed posse in Honey Grove, forty miles east of Denison, attempted to arrest a man for suspicion of murder. They pulled him from his cabin, shot him, then dragged his body to his church and burned it as other worshippers looked on.[17]

In these uncertain times, Easterly knew, it didn't take much to turn a crowd into a deadly mob.

The young reporter had plenty of opportunity to listen to the crowd. He spent most nights in early July at the promenade.

"As a transportation medium the bridge thus far has been about as useful as a whipsocket on an automobile," Easterly wrote, "but it is serving one public purpose anyhow—it is something of a social center for North Texas and Southern Oklahoma citizens. The concrete highway leading to the span in Texas and Oklahoma is a pleasant drive for a summer evening jaunt, and the scenery from the bridge is equally pleasing."[18]

Every evening, men and women from both sides of the river walked past the sign and wooden barrier and onto the completed span. A hundred individuals, sometimes more, strolled the bridge each night. Some were couples enjoying a night under the stars; others came to meet old friends. Some were families, welcoming a free night out of the house; others came simply to admire the pristine bridge.

"Those visiting the structure for the first time are impressed by the beauty of the steel and concrete span and evidence their surprise in enthusiastic praise."

To stand on the bridge *was* impressive. At 1,240 feet long, the bridge was constructed of four Pratt through-truss sections in a Parker camelback style. The steel chords, struts, posts, and bracing formed four arched latticework sections left and right, with more steel lattice above for a vertical clearance of a little over fourteen feet. The roadway was twenty-two feet wide, plus two twenty-inch safety shoulders. (The toll bridge, by contrast, was just sixteen feet wide in total, built at a time when speeds were slower and cars smaller.) The road surface of the new bridge was concrete over a solid flooring of steel beams and girders.[19]

The steelwork wore fresh coats of silver paint; the concrete deck was still unmarked by tires or fluids. At night, the bridge glowed crystal white like the Red River's own White City, illuminated by the full moon and the reflected light from gas spotlights beneath the bridge.

"An unusual psychology is manifest among those meeting at the bridge. It seems that a mutual regret or grievance provides all with an understanding that the formalities of an introduction often are not necessary."[20]

It was forty feet from the bridge railing to the mean river flow level below. The spotlights drew alligator gar, and they thrashed and rolled on the surface of red water that was viscous as silt. Even from forty feet above, there was the soft sough of the river below as it scraped over the sand bars. Occasionally, walkers would hear the wooden *tonk* of a floating tree trunk striking one of the concrete piers.

"From the watchman it is learned that there are a number of Durant and Denison citizens who rarely miss a night at the bridge."

Looking southward from the bridge toward Denison, the nighttime sky was soapy white from the lights of the rail yard. To the north, the sky was black; the communities on the north bank had no electricity, and their day ended at sundown. Stores with umbrellas and millinery and hair ribbons were to the south; the farms with fresh milk and garden tomatoes and smoked bacon were to the north. Oklahomans and Texans stood on the bridge that was to bring both sides of the river together and stared down-river a half-mile at the lights of the toll bridge and the automobiles that stopped to pay seventy-five cents each time they crossed it.

Easterly described two strangers who had stopped to admire the bridge, standing next to one another, leaning against the railing. "Each knows the other is there to view the new structure and to attest his displeasure that the span remains closed."

A layer of anger was building atop the dread that already blanketed the communities along the Red River.

By mid-July, Alfalfa Bill Murray had reached his boiling point. The Red River Bridge Company continued collecting tolls from thousands of travelers daily, and the new bridge—paid for with public money from Oklahomans and Texans—sat idle.

Ross Sterling appeared to have no ideas about how to solve the problem. His strategy seemed to be to hunker down and see what happened.

The continuing collection of tolls was sucking the financial lifeblood from farmers, dairymen, businesses, and families near the Red River, and Governor Murray couldn't understand how Governor Sterling would allow Texas sovereignty to be held hostage to a federal court judge.

"I'm reminded about the story of two Texas farmers who bought a cow," Murray told a group of friends when asked about the situation on the Red River. "A question arose as to how they would divide their interests. Ike claimed the back half of the cow; Moses was forced to content himself with the front.[21]

"Moses was pretty much bothered by the deal. 'I have to feed my half,' he complained, 'while Ike gets the milk from his half. After a while I get tired of the arrangement. The wife and I need some meat. So I killed my half and Ike's half died.'"

By mid-July, Murray decided it was time for some meat.

On Thursday morning, July 16, Governor Murray dictated to his office secretary a one-sentence letter. It was addressed to the Oklahoma State Highway Commission, and Murray asked his secretary to make four copies, all originals. Murray summoned Oklahoma Highway Commission chairman Sam R. Hawks—who some accused of being the governor's rubber stamp—to his office and, with some discussion, gave Hawks three of the letters.[22]

At 3:45 p.m., Division Engineer Burt McCaleb, of the Oklahoma Highway Department, drove south from Durant toward the Red River. McCaleb carried two copies of Murray's letter. With McCaleb in the car were three other department engineers; a truck with a work crew followed behind.[23]

Arriving at the Oklahoma end of the new free bridge, the engineers paused long enough to move aside the sawhorses blocking entry, then drove to the south end of the bridge and the watchman who was squinting back at them there. As the other three Oklahoma engineers retrieved tools from the trunk of the car, McCaleb handed one of Murray's letters to the watchman and asked him to read it.

"You are hereby ordered and directed that the state highway department is to proceed without further delay to open all three bridges over Red River," the letter declared, "and the Commission is hereby authorized to do all things necessary to remove obstructions from bridges, making free passage to each and all of them." The letter carried the typewritten signature of Gov. William H. Murray.[24]

As the watchman looked on, Burt McCaleb, the other engineers, and a crew from the Oklahoma work truck dismantled the barriers blocking the south end of the bridge. Workmen tossed the four-by-eight painted warning sign onto the gravel shoulder of the approach road, opening the free bridge to anyone who wanted to drive over it.

The first vehicle to cross the bridge (not counting McCaleb's highway department auto) was a Texas sedan from Denison with four people on their way to Durant. Word of the bridge opening soon spread with "cyclonic velocity," according to newsman Claud Easterly. Neighbors phoned neighbors with the news; business owners told customers; and a lodge meeting in progress was abruptly canceled when members ran out of the hall, leapt into automobiles, and roared off toward the river, eager to be among the first to cross the structure.[25]

Easterly's editor sent him to the bridge to survey the drivers. A check from 8:00 to 8:45 p.m. showed that nearly five hundred vehicles of all types—autos, trucks, buses, bicycles, motorcycles, and wagons—crossed the bridge, half from Oklahoma, half from Texas.

Almost every driver and passenger indicated their support, Easterly wrote, by waving, honking, or flashing lights. "Everybody smiled broadly. Many shouted, 'Hurrah for Governor Murray,' or 'Hurray for Alfalfa Bill.'" Those shouts came chiefly from Oklahomans, he said, but Texans weren't long in joining in.[26]

Ross Sterling was unaware of the shouts and auto horns and general hullabaloo resulting from the bridge opening. He had left his office early that afternoon to attend an event near Mineral Wells, two hundred miles away from Austin. The cavalry brigade of the Texas National Guard was ending its two-week summer camp with a formal review of troops by its commander in chief, the governor of Texas. The annual weekend event was more a men's smoker for the highest-ranking officers, political movers, and the governor. As Oklahoma highway workmen tore up barricades at the Denison bridge, Governor Sterling—unaware—was sitting down to a welcome evening of steaks, speeches, cigars, and ribald stories.[27]

Earlier that afternoon—just as McCaleb and his work crew were leaving Durant to open the bridge—Murray sent a telegram to Governor Sterling: "I have just issued an executive order opening the Denison bridge on Red River and instructed the engineer to tear up the roadbed on the Oklahoma side approaching the other bridge. Trust you will join me on your side of the line. It is absurd for any individual person or corporation to hold up two sovereign states and the project of the nation itself. Urge that you give like instructions on your side of the line."[28]

The wire was received at Capitol Station in Austin at 4:22 p.m. and was logged in at the governor's office just minutes later. Pat Dougherty, the governor's executive secretary, read the wire and attempted to reach Sterling at Camp Wolters, even as he fielded calls from highway commission members, the attorney general's office, and newspapers across the state. About 9 p.m., Dougherty got through to the governor and read him the text of Murray's telegram.

Sterling was flabbergasted that Murray would put him on the spot like this. How could he send state employees into Texas? How dare he ignore a lawful federal court order? Yet, underneath Sterling's astonishment, was a worry that Murray's involvement would further entrap him in that tar baby of a contract. Judge Kennerly's order held Sterling personally responsible for prohibiting travel over the free bridge. If the bridge remained open, Sterling could be cited for contempt of court. His attorney general was telling Sterling that the contract with the Red River Bridge Company was of dubious legality; Sterling could be held personally liable if the bridge issue didn't just go away. He *had* to shut down traffic on the free bridge right away.

Sterling needed a force on the ground in Denison, a force strong enough to take back the bridge, close it down, and see that it stayed closed. That shovel-leaning excuse for a watchman in Denison would be no help at all in retaking the bridge, and Sterling knew he couldn't rely on the county sheriff for help. (The sheriff was in one of those cars zipping back and forth across the now-open bridge.) Fortunately, the governor was right in the middle of a thousand armed and trained troops under his command. The Fifty-Sixth Cavalry Brigade of the Texas National Guard was wrapping up summer camp and was bivouacked right there at Camp Wolters. (The governor had been playing cards with the commanding general until he was interrupted by his secretary with news from Denison.) In fact, the 112th Regiment had extensive martial law experience from the Longview riots in 1918, Galveston strike duty in 1920, Borger martial law in 1928, and the Sherman courthouse riot the previous summer. They were just what the governor was looking for.[29]

The Texas state constitution names the governor commander in chief of the state militia, empowered to call on it to execute the laws, suppress insurrections, and repel invasions. That fit the bill, to Sterling's way of thinking. First, however, Sterling would have to consult with the adjutant general, the governor's aide in matters related to supervising the military department of the state. William W. "Bill" Sterling (no relation to the governor) was adjutant general, responsible for the Texas National Guard and the Texas Rangers.

Fortunately, Bill Sterling was right there at Camp Wolters, also attending the review of troops. The governor sent word to fetch his military adviser.

As Governor Sterling fretted about events at the Denison bridge, Governor Murray was back in his office after supper in the residence with his wife. (Murray found having a state-paid cook was wasteful, and he liked his wife's cooking better anyway.) A caller tipped Alfalfa Bill that the Red River Bridge Company officers would seek a restraining order the next morning in Durant district court to prevent Murray from tearing up the toll road approaches.

Thursday, July 16, 1931, the first day of what newspaper reporters would come to call "The Red River Bridge War," ended with Governor Murray, still in his office with feet on his desk, making calls to a few old friends in Durant. Governor Sterling sat at a desk in a suite at the Baker Hotel in Mineral Wells near the encampment at Camp Wolters, conferring with his adjutant general about how to respond to the Oklahoma incursion. At the Red River, between Denison and Durant, the new steel bridge gleamed white under the moonlight. Most of the earlier joyriders were at home and in bed, but late-night cars and trucks and buses ran north and south over the bridge, never slowing to pay a toll, their big tires singing high notes on the new pavement.

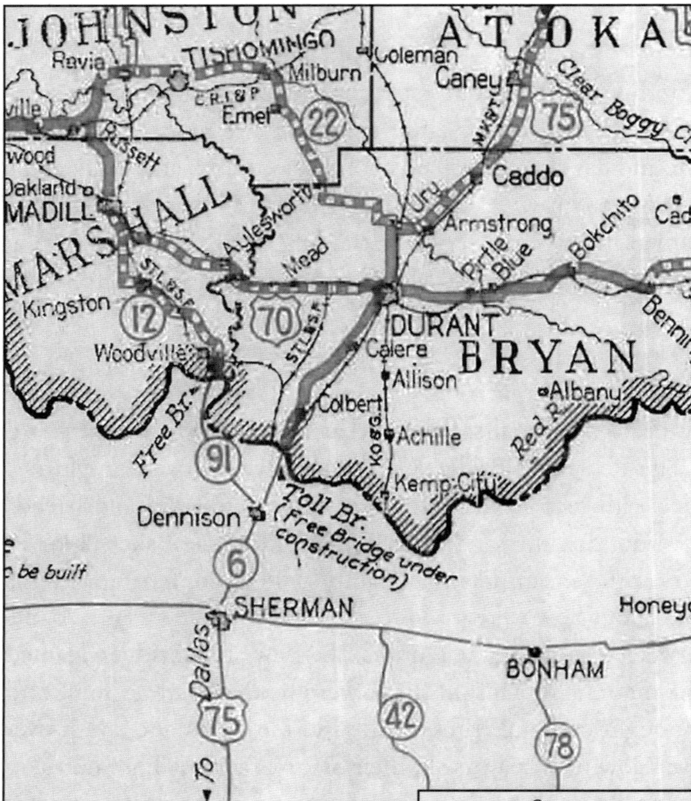

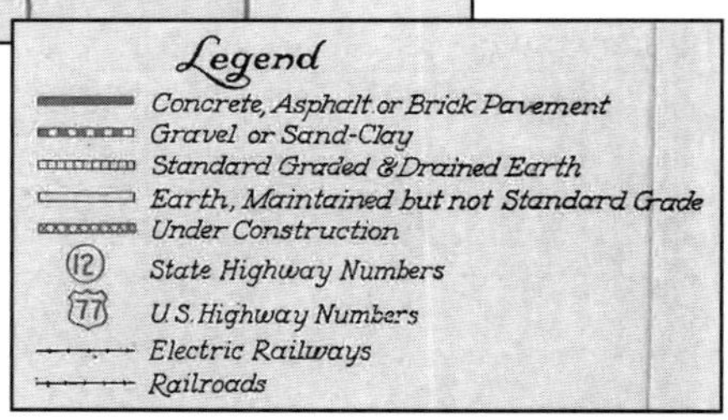

This 1930 Oklahoma highway map shows an improved highway running between Durant and Dennison [sic]. As the map was printed, the new toll-free bridge was under construction less than a half mile upstream from the "Toll Br." Some motorists used state Highways 91 and 12 to detour around the barricaded bridges, adding an extra twenty-one miles over gravel road and a harrowing river crossing on an ancient plank bridge. (Courtesy of the Oklahoma Department of Transportation)

Frank Colbert's workers used stout poles to propel the ferry from one riverbank to the other. Strong currents could send the ferry out of control; fractious livestock could overturn it. (Heye Foundation Collection, Courtesy of the Oklahoma Historical Society, 12621-A)

This gasoline-powered ferry was faster than the poled ferries and could carry heavier loads, such as the automobiles and farm trucks wanting to cross the Red River at the old Colbert crossing. (DeGolyer Library, SMU, Texas Photographs, Manuscripts & Imprints, ag2 005.0001)

The *New York Times* referred to Frank Colbert as one of the wealthiest men in Indian Territory. Colbert built his family home, Riverside, from the proceeds of his Red River ferry franchise. (Robert L. Williams Collection, Courtesy of the Oklahoma Historical Society, 6755)

As late as 1920, many Oklahoma and Texas farm roads and country bridges were primitive pathways, unsuited for dependable travel by automobile or truck. Both states launched aggressive road-building programs. (Courtesy of Texas State Library and Archives Commission)

Frank Colbert and other investors chartered the Red River Bridge Company in 1890 for the purpose of building a steel toll bridge. Completed in 1891, the bridge was washed away during the flood of 1908 and rebuilt in 1915. (From the Collection of Richard Hood)

Herb Fielder and his passenger pose in 1920 on the Red River Bridge Company's new auto bridge. The men were on a cross-country trip and were equipped to navigate the nation's primitive highways. (Courtesy of Tyrrell Historical Library, Beaumont, Texas, 2010.078.228A)

After spring rains, unimproved roads could become muddy quagmires. This driver relied on mud chains, but they were not enough to prevent him from becoming bogged down to the axles. (Courtesy of Tyrrell Historical Library, Beaumont, Texas, 2010.078.048A)

Even the runoff from a light rain or a nearby spring could leave water on roads that had been plowed flat, leading to dangerous skids, becoming stuck in mud, or roadway washouts. (Courtesy of Tyrrell Historical Library, Beaumont, Texas, 2010.078.149A)

Texas governor Dan Moody (*left*) appointed businessman Ross S. Sterling (*right*) chairman of the Texas Highway Commission in 1926. As chairman, Sterling entered into a questionable contract expediting construction of a new Red River Bridge. (R. S. Sterling Collection, Wallisville Heritage Park)

As a first-term Texas state representative, J. J. "Jake" Loy introduced the bill that made the new free bridge possible. Three years later, as a state senator, he introduced the bill that ended the bridge standoff. (Forty-First House Composite; Detail: J. J. Loy; Accession ID: CHA #1989.539; Courtesy State Preservation Board; Austin; TX, Photographer: Jesse Herrera; 8/18/95; post conservation)

Former US senator Joseph W. Bailey was counsel to the Red River Bridge Company until his sudden death in a Sherman courtroom. Son "Little Joe" Bailey took over his father's bridge cases. (US Senate Historical Office)

William H. Murray arrived in Tishomingo, Indian Territory, in 1898. There he found a wife and began a family while building a lucrative law practice in the Chickasaw Nation. (Juanita Johnston Smith Collection, Courtesy of the Oklahoma Historical Society, 6985)

A federal court ruling closed the new Durant-to-Denison toll-free bridge before it could open. The bridge remained closed for three weeks, backing up traffic for miles and cutting off trade between Texas and Oklahoma. (Courtesy of Texas State Library and Archives Commission)

Oklahoma governor Murray bursts through the federal injunction to open the new toll-free bridge in this editorial cartoon, which appeared in both the *Dallas Morning News* and the *Daily Oklahoman.* (Reprinted with permission of the *Dallas Morning News*)

Texas Rangers erect a barricade, preventing anyone from crossing the new bridge. Adjutant General W. W. Sterling (with rifle) stands in front of the sign; Capt. Tom Hicks stands to the left. (Stan Oliner Collection, Courtesy of the Oklahoma Historical Society, 20381.10)

Local residents, out-of-towners, and even airplane sightseers came to the Red River day and night to gaze at the empty bridge. For some, a toll-free bridge could make the difference between poverty and prosperity. (Courtesy of Texas State Library and Archives Commission)

A semi-literate country boy, Ross S. Sterling pulled himself up by his bootstraps to become founder of Humble Oil and Refining (predecessor to Exxon/Mobile) and, eventually, governor of Texas. (Courtesy of Texas State Library and Archives Commission)

National Guard rifles, a stout barricade, and two hundred yards of plowed and impassable roadway met motorists attempting to reach the toll bridge from Oklahoma. (Murray allowed mail carriers free passage, however.) (Stan Oliner Collection, Courtesy of the Oklahoma Historical Society, 20381.13)

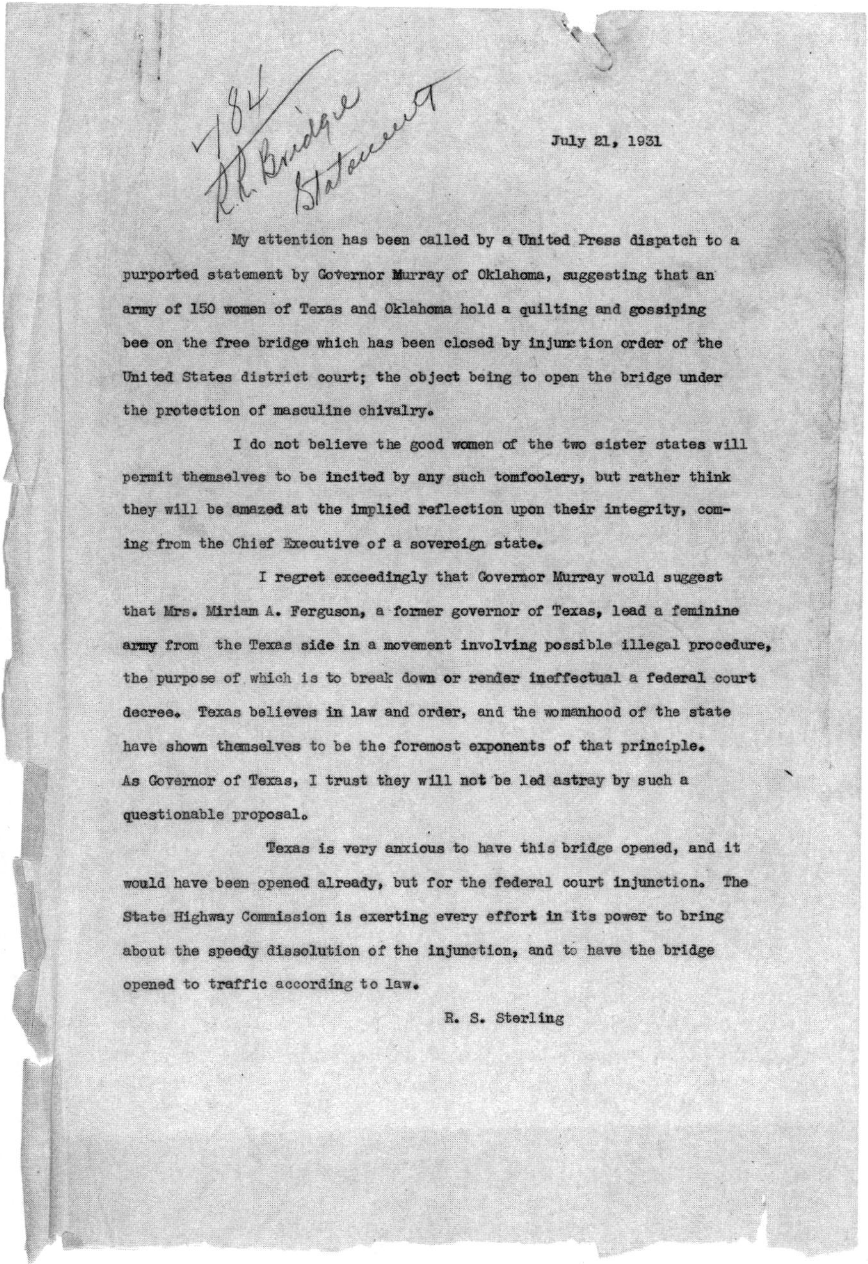

784
R.R. Bridge
Statement

July 21, 1931

My attention has been called by a United Press dispatch to a purported statement by Governor Murray of Oklahoma, suggesting that an army of 150 women of Texas and Oklahoma hold a quilting and gossiping bee on the free bridge which has been closed by injunction order of the United States district court; the object being to open the bridge under the protection of masculine chivalry.

I do not believe the good women of the two sister states will permit themselves to be incited by any such tomfoolery, but rather think they will be amazed at the implied reflection upon their integrity, coming from the Chief Executive of a sovereign state.

I regret exceedingly that Governor Murray would suggest that Mrs. Miriam A. Ferguson, a former governor of Texas, lead a feminine army from the Texas side in a movement involving possible illegal procedure, the purpose of which is to break down or render ineffectual a federal court decree. Texas believes in law and order, and the womanhood of the state have shown themselves to be the foremost exponents of that principle. As Governor of Texas, I trust they will not be led astray by such a questionable proposal.

Texas is very anxious to have this bridge opened, and it would have been opened already, but for the federal court injunction. The State Highway Commission is exerting every effort in its power to bring about the speedy dissolution of the injunction, and to have the bridge opened to traffic according to law.

R. S. Sterling

Governor Sterling failed to see the humor in Murray's quilting bee suggestion. In a statement to the press, he said he doubted that the women of the two states would be incited by such "tomfoolery." (Courtesy of Texas State Library and Archives Commission)

Seated in a war-surplus camp chair and wearing a campaign hat, Alfalfa Bill answers reporters' questions while visiting the "war zone." Note Murray's ever-present cigar and the newspaper in his left jacket pocket. (Courtesy of the Oklahoma Historical Society, 2012.201.B0409.0484)

Murray had a publicist's eye for a photo opportunity. Here, he waits for his escort column of guardsmen to catch up before inspecting the battle lines for the benefit of the newsreel cameraman. (Courtesy of the Oklahoma Historical Society, 2012.201.B0409.0494)

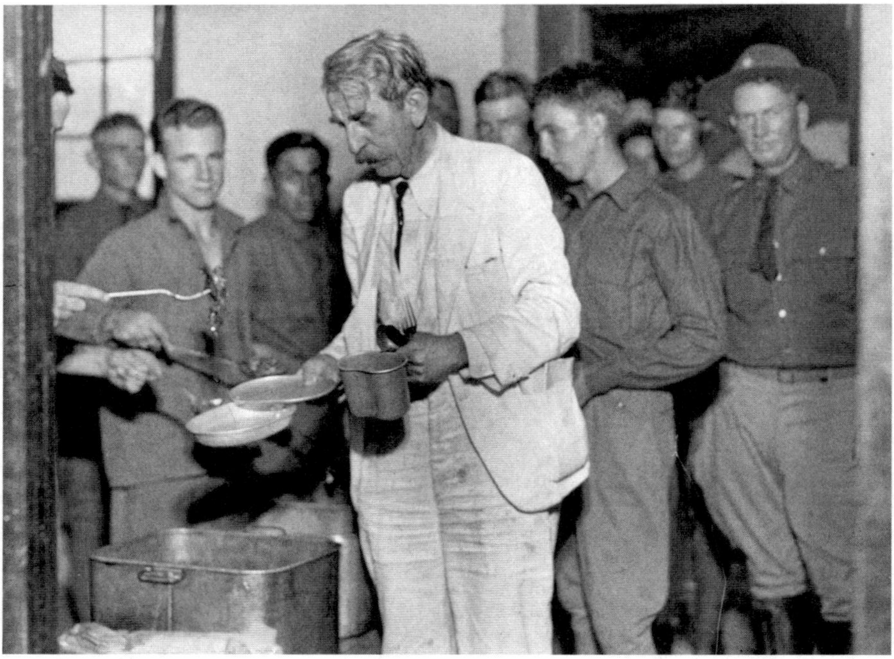

A rumpled Alfalfa Bill shares an evening meal of liver and onions with his troops at the Red River encampment. Never one for ceremony, Murray stood in the chow line with the guardsmen. (Courtesy of the Oklahoma Historical Society, 2012.201.B0409.0490)

"The free bridge is open now," Alfalfa Bill told reporters at breakfast. "If people are fools enough to want to pay seventy-five cents to cross here, let 'em do it." (Courtesy of the Oklahoma Historical Society, 2012.201.B0409.0489)

Alfalfa Bill poses with Lt. Col. John MacDonald, commander of troops at the bridges. MacDonald refused service of a federal court order, saying he took orders only from Governor Murray. (Courtesy of the Oklahoma Historical Society, 2012.201.B0409.0499)

The situation may have been tense on the Red River front, according to this editorial cartoon, but not particularly serious. Most newspapers overlooked the serious consequences of the bridge closures. (Reprinted with permission of the *Dallas Morning News*)

Armed Oklahoma National Guardsmen turn away a man attempting to walk across the toll bridge. (Note the flag used to mark the limits of Governor Murray's martial law zone.) (Courtesy of the Oklahoma Historical Society, 2012.201.B0409.0493)

When completed in 1931, Texas and Oklahoma's new steel truss bridge was state-of-the-bridge-builder's-art. It would carry up to twenty thousand vehicles a day across the Red River for the next sixty-five years. (From the Collection of Richard Hood)

His actions during the Red River Bridge War earned Governor Murray a nationwide reputation, and as candidate for the Democratic presidential nomination, he drew large crowds wherever he appeared. (John Dunning Political Collection, Courtesy of the Oklahoma Historical Society, 23139.G-12)

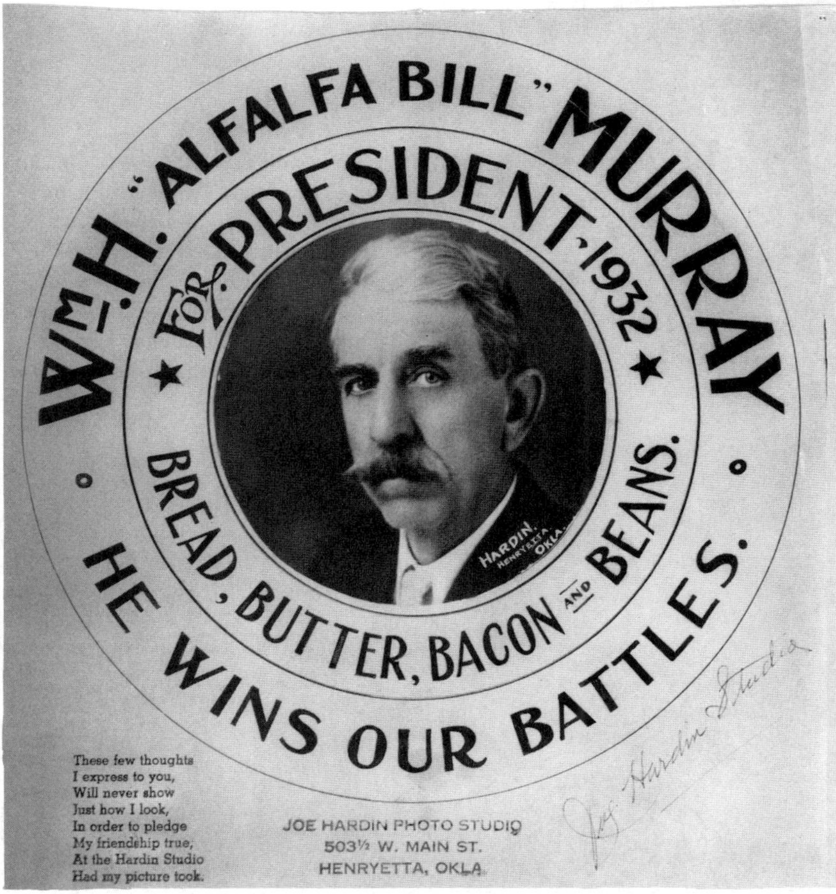

Alfalfa Bill's campaign slogan of "Bread, Butter, Bacon, and Beans" resonated with Depression-pinched rural voters who had seen crop and livestock prices fall even as the cost of staples increased. (John Dunning Political Collection, Courtesy of the Oklahoma Historical Society, 23139.G-11)

Murray won a second term as Oklahoma governor and served until January 1935. During his administration, he declared martial law or called up the National Guard more than two dozen times. (Barney Hillerman Collection, Courtesy of the Oklahoma Historical Society, 21412.M49.2)

Increasing traffic across the Red River led Texas and Oklahoma in 1958 to cooperate once again in building a companion to the 1931 free bridge. One bridge would carry northbound traffic, the other southbound travelers. (Courtesy of the Oklahoma Historical Society, 2012.201.B0106.0783)

CHAPTER 8

Midnight Bulldozers

AT DAWN the following morning the new bridge was quiet, the only sound the hiss of the river over its sandy bed and the rustle of wind in the cottonwood trees that lined the river on its south bank. No tires sang high notes on the pavement; no vehicles passed north or south.

On the downslope of the Texas approach road, just before the concrete that marked the bridge itself, were two sturdy wooden barricades stretched across the highway, the four-by-eight painted "Warning!" sign, and two armed Texas Rangers there to halt any traffic attempting to cross the Red River on the new free bridge.[1]

After some dithering the night before (and with strong advice from his adjutant general) Gov. Ross Sterling had decided to hold off sending armed troops to close the bridge. Adjutant Gen. Bill Sterling, a ranger himself, assured the governor that two Texas Rangers could do as much as Texas National Guardsmen for the moment, and, if the elite law enforcement unit needed help, they could always call out troops later.[2]

With Governor Sterling's approval, General Sterling arranged to meet up with Texas Ranger captain Tom Hickman. Together, they drove to Denison and scouted the situation before dawn.

The Texas Rangers have a history that stretches back to before the Republic of Texas days. As early as 1823, impresario Stephen F. Austin hired "rangers" to protect his fledgling Texas colony from Indians and other predators. Lacking any other local law enforcement, presidents of the Republic of Texas hired men to "range" the territory and keep the peace. As Texas' population swelled after the Civil War, settlers pressed the western boundaries of the new state. Indians pressed back, often by burning ranches, stealing livestock, scalping settlers, and kidnapping white children. The governor's Texas Rangers—extremely hard, extremely experienced men—were assigned to protect settlers by chasing Indians and killing them when they could. The rangers were heroes to early Texas residents and legends throughout the West. By the twentieth century, with marauding Indians gone from Texas, the rangers were dealing with oil patch claim jumpers, border banditry, rustling rings, serial bank robbers, and any other issues the governor felt transcended the capabilities of jurisdictional law enforcement. Texas Rangers were more than a state police force; they were legendary, an elite policing unit organized along military lines, under the direction of the state's adjutant general and the control of the Texas governor.[3]

Gen. W. W. "Bill" Sterling looked like a Texas Ranger ought to look: six feet four inches tall (a little taller in boots) with broad shoulders, narrow hips that allowed a gun belt to hang low, piercing blue eyes, and the rugged good looks of a cowboy movie star. (He was chosen by sculptor Gustave Borglum as the model for the proposed Texas Ranger monument intended for the state capitol grounds.) He had a reputation for keeping his mouth shut, saying little and doing much. Sterling was born in 1891 and raised in the mesquite and chaparral borderlands of deep South Texas. As a boy, he often visited the ranger camp near his family's ranch and, by fifteen years old, made it known that he wanted to be a Texas Ranger. First, though, he attended Texas A&M, then served as chief scout for the US Army during the Mexican borders war of 1915. Turned down for regular military service because of an earlier wound, he joined the rangers and rose to the rank of captain. Before being appointed adjutant general, Sterling commanded a troop responsible for patrolling about a third of the long Texas-Mexico border.[4]

Bill Sterling and Tom Hickman drove through the night from Mineral Wells to reach Denison, arriving well before dawn. Several Texas Highway Department employees were at the bridge on standby, and the rangers directed them to build new barriers and place them at the entry to the free bridge. As the highway workers completed the barriers, Sterling and Hickman retrieved the large "Warning!" sign the Oklahomans had thrown into the ditch the day before and asked the workers to affix the sign to the completed barrier.

The bridge was closed, and there was some minor confusion at first—southbound vehicles driving onto the bridge, only to have the rangers turn them back, and Texas motorists driving up to the free bridge, thinking it was open—but by daybreak the free bridge was empty of all traffic, and motorists were once again crossing the Red River on the old toll bridge and paying seventy-five cents for every crossing.[5]

An early-rising Governor Sterling finally completed his response to Governor Murray. If anything, his sense of indignation had risen since the night before.

"Understand your highway department has crossed into Texas and torn up approaches to the bridge," he wrote. "I feel you have extended your jurisdiction beyond all reason. I cannot join you in making federal court injunction inoperative. At request of Texas Highway Commission I have sent state rangers to protect Texas highway employees while they are endeavoring to obey the federal court's injunction."[6]

Sterling phoned his office and dictated the message to his executive secretary, who would send it by wire to Governor Murray's office. Sterling called for his driver to take him from the hotel to Camp Wolters, where he would sit in the grandstand watching field day events. That afternoon he would ride in the closing parade.[7]

The governor would be difficult to reach with updates on the bridge situation, but he couldn't think of anything more that could happen.

Also an early riser, Alfalfa Bill was in his office and at work by seven o'clock. It was a Friday, a work day, and he intended that the citizens of Oklahoma get their money's worth (though, in a phone call the night before, he likely encouraged a Durant friend, also a state judge, to take the day off and go fishing).[8]

Governor Murray spent much of the morning returning calls from newspapermen. He knew from experience that newsmen are like good farm dogs: feed them well and they'll hunt for you. The big-city newspapers—Oklahoma City, Tulsa—would never be his friend, but Murray nurtured the support of the small-town editors and publishers. He would check in on the Oklahoma papers later; his first calls were to newspapers in Texas.

To the *Denison Herald* he explained why he sent highway department men to tear down the barricades. "We have a right to cross a state boundary under United States laws. When we cross a bridge and find an obstruction, we have the right to tear it out of the way." Did you expect any resistance?

"There was an old watchman there and I instructed that he not be hurt. I instructed highway department officials to take him down and take his pocket knife and chewing tobacco away from him."[9]

The *Sherman Democrat* was next. The reporter asked why Murray chose to open the free bridge. "I assumed that the Texas government wanted the Denison bridge opened, otherwise it wouldn't have been so foolish as to create an expense to complete the highway, then close it." Asked if he had plowed up the Oklahoma entrance to the toll bridge, Murray said his highway men were waiting for a bulldozer to arrive at the site. Meanwhile, he had posted a man at the toll bridge entrance to direct traffic westward to the Preston Crossing, a combination high-water crossing and single-lane bridge twenty miles upriver. "I call on the citizenship of both states to take a little trouble and go by the Preston bridge."[10]

Murray's call to the *Austin Statesman* was intended to twist Governor Sterling's tail in the Texas capital city. At the time of the call, Murray had not yet received Governor Sterling's telegram, so the reporter read it to Alfalfa Bill, starting with "you have extended your jurisdiction beyond all reason." From that point on, the reporter said, "the interview became a monologue with Governor Bill doing the honors."[11]

"Now, listen and get my point," Murray interrupted. "Oklahoma owns half of that bridge lengthwise. A bridge that just goes to the middle of the stream is no bridge at all." He then launched into a lecture on the Adams-Onís Treaty.

Murray expressed disappointment over Governor Sterling's failure to join him in the forcible opening of the bridge to allow free traffic between Texas and Oklahoma. "Since your governor doesn't want it open, he has the power to keep his side closed. However, I predict that in a few days he'll stand alone in the matter. He won't have public support with him."

"If Governor Sterling had just kept his hands off the bridge, it would've been opened," Murray said. Still wound up, Alfalfa Bill charged that Governor Sterling "is just trying to make Oklahoma pay half," of a $60,000 settlement demanded by the toll bridge company.

The lecture lasted a half-hour before Murray rang off, pleading the press of state business.

As Alfalfa Bill was speaking with the *Austin Statesman*, state senator Jake Loy was in his Austin legislative office weighing a difficult political decision. Loy was not yet a skilled politician, but he knew that there would be consequences for going up against a sitting governor of his own party. Still, there was just something not right about Governor Sterling's role in all this. Also, he knew his constituents were being mistreated, still shackled to a toll bridge and denied the use of a newer, safer free bridge for which they had already paid.

That afternoon Senator Loy announced that on Monday he would go before the senate and call for a legislative investigation to determine why the bridge was not open, why the Texas Highway Commission entered into the contract of dubious legality as it did, and why the governor should put the state of Texas to the expense of enforcing a federal injunction. He gave a copy of his announcement to the wire service reporters in the capitol pressroom for distribution to newspapers and radio stations throughout the state.[12]

Loy's next task was to reach into a desk drawer for a blank Western Union form. He drafted his wire to Governor Murray: "As author of the Texas free bridge law I want to congratulate you on your courage in opening the free bridge at Denison, which has been unnecessarily delayed by technicalities. I hope our governor will make every effort to cooperate with you in opening the free bridge for public use."[13]

The young state senator made sure Sterling's office received a copy of the telegram.

Joe W. Bailey Jr. made an early morning drive to Durant to meet with his clients, the owners and officers of the Red River Bridge Company. The attorney advised A. G. Gilbert, a company shareholder from Colbert, to file suit preventing Governor Murray and the Oklahoma Highway Department from blocking access to the toll bridge.

Bailey drew up the petition and delivered it to Judge Porter Newman's district court by nine o'clock that morning, planning to ask the judge for an immediate hearing. Strangely enough, Judge Porter hadn't arrived at his courtroom; court officials could say only that the judge would be out of the building all day. The judge couldn't be located at the coffee shop, the barbershop, or any of his other usual stops.[14]

By late afternoon, the judge was still absent and Bailey drove to his home. No, the judge isn't home, his wife said. He left early this morning for an unexpected fishing trip.

Bailey had little choice but to park outside the judge's home, hoping he wouldn't be "fishing" all weekend.

"Eating and sleeping are not considered when duty calls for a newspaperman to be on the job and furnish the readers with facts as fast as they develop." That was the published expectation at the *Denison Herald*, and reporter Claud Easterly was living up to it. After a late night watching cars race across the free bridge, Easterly was up, dressed, and back at the bridge before breakfast Friday morning to see the Texas Rangers pull the plywood "Warning!" sign out of the ditch.[15]

Easterly didn't know Gen. Bill Sterling, but he had a nodding acquaintance with Tom Hickman from the time the ranger testified at the Sherman riot trial the previous summer. Sterling was all spit-and-polish, standing at a modified parade rest and wearing his "Sunday gun," a pearl-handled Colt revolver he carried mainly for show. General Sterling declined to comment about what orders Governor Sterling had given them, other than to close the free bridge and keep it closed.

Tom Hickman was a spindle-shanked former rodeo rider described as having a smile as broad as his Stetson. For seven outlaws (to date) that smile was the last thing on earth they would ever see. Hickman knew (and could tell) every campfire tale that had ever been told. He also had a reputation for speaking freely with the press, and Hickman would do so with Easterly for the next few weeks. For the moment, however, he said only that they were expecting two more Texas Rangers to arrive later that morning. Jim Huddleston and W. H. Kirby would be driving up from Dallas with the supply trailer and camping gear. That meant, Hickman said, that the rangers would probably be at the bridge for some time.[16]

Easterly drove to the toll bridge and crossed the river (paying his seventy-five cents, of course). He drove a half-mile north but was stopped before he came to the main highway. There were no armed guards at the flimsy-looking barricade there, but an Oklahoma Highway Department worker said, "You'll have to run over a man or so to get by." Highway workers were preventing Oklahoma traffic from going south across the toll bridge and directing northbound motorists to turn around and return to Denison. (The toll taker charged another seventy-five cents for drivers who were turned back across the river.)[17]

The *Herald*, an afternoon newspaper, had a deadline of three o'clock. When Easterly returned to the newsroom, it was as chaotic as he had ever seen it. "The *Herald* was besieged by telephone calls from many North Texas and Southern Oklahoma newspapers wanting facts about the incident," he said. "Calls also came from eastern and Midwestern points."[18]

Newspapers from New York and Chicago picked up the story of Governor Murray's seizure of an interstate bridge from the wire services and, without even knowing all the details, were considering dispatching reporters to Denison. Hearst Metrotone News, a newsreel company, was already making arrangements to get bridge footage.

Many on the *Herald* staff stayed at the office in the afternoon to read the first edition as it came off the press. The bridge coverage was extensive, and those headlines probably bumped from the front page two stories that, on a slower news day, would have been more prominent.

The first story described plans for the US Highway 75 Association Caravan, scheduled to pass through Denison later that month. The old Winnipeg-to-Galveston King of Trails Highway was now US Hwy. 75, and as the longest north-south completed highway in the nation was a major tourist artery that meant a lot to Denison, Sherman, Durant, and other towns along the route. More than a hundred autos would depart Galveston midweek and, picking up other travelers along the way, were expected to stop over in Denison before crossing the Red River on their way to Canada.[19]

Another story that might have gone unnoticed was a report by the Bureau of Labor Statistics saying employment had declined nationwide in June; unemployment was now at its highest recorded level. President Hoover assured the nation that private relief agencies were capable of assisting any who needed it, and he insisted the economy was already turning around. Nevertheless, an increasing number of families were walking the knife-edge between fear and anger.[20]

Friday's *Herald* carried a front-page editorial summing up initial reaction to the bridge fracas and the general mood of the community. "The sudden and climactic development in the free bridge controversy has left the people somewhat stunned and wondering what to expect next," the editorial said. "Public opinion is aroused and people are clamoring for a final decision. Meanwhile, state officials haggle over contracts, exert their prerogatives, and keep the public in suspense."[21]

Easterly passed up supper to take one more drive out to the bridge.

At the end of the day, as Governor Sterling sat down at a formal awards dinner at Camp Wolters, Governor Murray left the capitol and walked back to his residence. Friday had been a good day for Alfalfa Bill.

Murray was a constitutional scholar and a states' rights proponent; he repeatedly told the newspapers that the Eleventh Amendment to the US Constitution precluded entangling a sovereign state in the federal suit of a private citizen. The federal injunction was a nullity, and Governor Sterling appeared to be holding the new bridge hostage, waiting for Oklahoma to bail him out of that contract with the bridge company.[22]

An opinion of Assistant US Attorney General Seth Richardson backed Murray's position. Neither Murray nor the state of Oklahoma could be bound by the federal injunction that bound Governor Sterling. Alfalfa Bill knew he could always call out his National Guard if Judge Kennerly came after him with federal marshals.[23]

Murray also avoided entanglement in an Oklahoma state court, too, thanks to Judge Newman's spontaneous fishing trip. Public opinion would open that bridge, Murray believed. He had the support of Oklahomans; Texans would be with him soon.

All in all, it was a good day for Alfalfa Bill.

Four Texas Rangers sat around a campfire laid among the spinous bushes and stunted trees alongside the road near the bridge barricade. At sundown they lit black melon-shaped oil lamps and set them across the highway as a warning, but there was no traffic. A heavy rain that afternoon passed by suppertime, but the rangers had rigged a canvas lean-to against the trailer that Huddleston had brought to Denison that morning.

The night was quiet except for the sound of Tom Hickman telling one of his long and pointless campfire stories. Near midnight, they heard the grumble and crunch of heavy equipment working on the far side of the river.

Burt McCaleb, Oklahoma highway engineer, finally found the bulldozers needed to plow up the toll bridge approach roads. By morning, the only way for a driver to reach the toll bridge from Oklahoma would be for someone to drive a ton-and-a-half automobile across two hundred yards of asphalt chunks and gummy soil.[24]

Travel over the Red River on US Hwy. 75 between Durant and Denison effectively ended late Friday night, July 17.

Joe W. Bailey Jr. arranged to meet Judge Porter Newman in his Durant courtroom at eight o'clock Saturday morning. Bailey presented his application for a restraining order to prevent Governor Murray and the Oklahoma Highway Department from blockading or doing further damage to the road leading to the toll bridge.

Judge Newman told Bailey he would take the matter under advisement and have a decision by one o'clock that afternoon.

Governor Sterling arrived at his office in the capitol at about eight o'clock after an early morning trip to Austin from Mineral Wells. In his absence, the governor's office had been flooded with telephone calls, and there was a message that Senator Jake Loy was at present in the building and wanted to confer with the governor.

Sometime during the morning, Sterling was shown (or told about) news coverage of events at the Red River. Every major newspaper in the state featured Murray's actions on their front pages, the stories often topped with headlines two inches high. The *New York Times*, *Washington Post*, and *Los Angeles Times* were carrying news about the "war" on the Red River that made Texans and Oklahomans look like bumpkins. The Austin newspaper—the one delivered daily to the Texas governor's mansion—ran the headline "Alfalfa Bill Leads Bridge Opening Fight." Editorial writers for Texas' largest Sunday newspapers harrumphed over Governor Murray's extralegal showboating, but there were no ringing endorsements of Sterling's inaction. The *Fort Worth Press* called the standoff "a nauseating legal mix up," and urged Sterling to settle up on the contract and open the bridge; the *Houston Post* groused that "in Texas we must wait until the lawyers quit wrangling before the people get their rights."[25]

Understandably, Governor Sterling was not in the best of moods when Jake Loy was shown into his office. Jake Loy had been under a lot of pressure the past forty-eight hours too. Traffic was being routed around Denison, and merchants were missing out on a big day of sales as well as their regular summer tourist trade. Loy's senatorial constituency also included some major Red River Bridge Company shareholders, and they were concerned that the state wasn't meeting its obligation to the bridge company. Loy wanted—needed—the Texas Rangers withdrawn and the free bridge opened right away. Short tempers may have accounted for the loud voices emanating from the governor's office and were heard by a reporter for the *Austin Statesman.*[26]

Loy later told the reporter he had informed Governor Sterling that in ordering rangers to protect the bridge, the governor's "executive authority had been exceeded." He said that because the free bridge had been closed by federal injunction, "keeping it closed was a matter for federal authorities," if that's what they wanted.

Despite the strong words (and hinting that the bridge owners might settle for a lower payment), Loy persuaded the governor that he must take some sort of action. Sterling agreed to meet with all parties to work out a plan for opening the free bridge. Assuming everyone could get to Austin, he agreed to convene a conference Sunday afternoon.[27]

After Loy left his office, Sterling dictated a terse note and asked his executive secretary to distribute it to the newspaper people. "I did not send rangers to Denison to close the bridge," he said, "but sent them there by request of the highway department to protect highway employees who were acting under a federal court order to keep this bridge closed."[28]

Jake Loy was generous in his comments about the governor when he sent a telegram to the Denison and Sherman newspapers that afternoon: "I have been in conference with Governor Sterling this morning trying to work out and settle the bridge controversy in Denison. The governor is doing everything in his power to work out a plan which will open this bridge legally at the earliest possible moment, and he has called the highway commissioners to meet with him and myself and the receiver of the Denison bridge company tomorrow here."[29]

At the Durant courtroom that afternoon, Joe W. Bailey Jr. picked up a written answer to his request to prevent Oklahoma highway workers from blocking access to the toll road. Judge Newman was succinct. He denied the restraining order, believing he had no authority to override the executive order issued by Governor Murray.[30]

(Later, at the coffee shop, the judge told the local newspaper editor he felt granting a restraining order would just prolong the whole bridge controversy.)

Bailey was forced to drive the extra miles to the detour crossing at Preston. With rangers and barricades blocking the free bridge, and with the approach road to the toll bridge plowed up, there would be no crossing the Red River between Durant and Denison.

Governor Murray spent the day in his office taking care of routine business, some of the time reading congratulatory telegrams. A group of businessmen formed the Carter County Murray Club and sent him their first resolution. "This meeting does heartily commend the action of Governor Murray," the resolution read, "for the Jacksonian manner in which he has attacked the toll bridge incubus." He sent a note to his attorney general instructing W. C. Lewis to prepare a petition on Monday that could be filed with the US Supreme Court.[31]

But Alfalfa Bill didn't think it would be necessary to go to Washington. He was convinced that Governor Sterling's position was untenable as well as unconstitutional, and that the governor must eventually capitulate. Already, newspapers were ready to declare Murray the winner; the headline in the *Daily Oklahoman* that morning was "Victory Near for Murray in Toll War." He told a reporter in the capitol hallway that he'd "been swamped with wires all day from Southern Oklahoma and all parts of Texas, even to the gulf coast." "There are more people in Texas who want these bridges opened than in Oklahoma," he added. Families along the Red River valley wanted free passage to the other side of the river; many of them *needed* free passage for better access to markets or mills or gins or wholesalers or jobs. Alfalfa Bill intended to open the free bridge, and he was well on his way to doing so.[32]

With one phone call, Governor Murray's attention shifted from the toll bridge problem to a much bigger problem, one that threatened to scorch the entire state. Cicero Murray, a distant cousin Murray had appointed to coordinate the state's jobless relief efforts, called to say that a mob of three hundred unemployed men and women were marching through the streets of Henryetta, demanding food at grocery stores and mercantiles and getting it. They had already collected about a hundred dollars' worth and were sharing it on the steps of Henryetta's city hall. So far, there had been no violence, Cicero Murray said, but the group threatened to come back to town every Saturday until they could find work.[33]

Governor Murray was fearful that an incident like this could be the spark that brought the tens of thousands of jobless Oklahomans into the streets of other towns across the state. Food riots were occurring weekly in cities across the country but, by and large, the press refrained from reporting them to avoid civic shame. Businesses had spiraled downward as a result of a prolonged drought, bottomed-out crop prices, bank failures, and foreclosures. All these things could result in despotism, revolution, or worse, Murray feared. The unemployment situation in Oklahoma was critical. The Red Cross provided emergency relief to which the state contributed, but available funds were long gone.

The governor called reporters into his office and then phoned the mayor of Henryetta. "The people are hungry and something must be done and will be done," Murray promised. He had collected several hundred dollars in a private relief fund by asking state employees to "voluntarily" contribute 10 percent of their paychecks. He would instruct Cicero Murray to turn that amount over to a Henryetta Relief Committee that the mayor was organizing.

The governor made it clear that he was no friend of the dole. "You can't just feed them or they'll sit down and expect it," he said. Murray asked the mayor to make a list of work that could be done, "such as cleaning up," and to meet with him at the capitol on Monday.[34]

Murray said he believed the people of all parts of the state realized from his speeches that they would receive aid if they needed it.

Attorney General James V. Allred of Texas was making it more difficult to solve the toll bridge issue. Shortly after Jake Loy and Governor Sterling announced the Sunday conference to negotiate a settlement to the bridge contract, Allred told the press his office was working on a legal opinion declaring the contract invalid and void. He said he would oppose any payment required for settlement with the Red River Bridge Company—a position likely to prolong the legal wrangling.[35]

Jake Loy got the news that afternoon while still in his capitol office. "The forthcoming opinion of the attorney general will add complications," he admitted, "the effect of which cannot now be forecast."[36]

To state senator Jake J. Loy, it seemed the settlement conference might be over before it ever began.

All weekend, Jake Loy received telegrams about the bridge issue, most from constituents and many containing well-meaning (if impractical) advice. One of the wires was from Judge B. F. Gafford, the sixty-year-old county attorney from Sherman who had once served as state representative from the area.

"Bridge matter is most serious to the public," he wrote. "Suggest you introduce resolution authorizing Red River Bridge Company to sue State for any sums due it under contract of purchase." If Loy's conference on the following day failed to produce a settlement, the former judge said, "I am sure that they may agree that during pending litigation the bridge may be open for public use."[37]

Loy glanced at the telegram and, not fully understanding the legal mechanics of Judge Gafford's suggestion, added the wire to the tall stack on his desk to be answered on Monday.

CHAPTER 9

Very Much Disturbed

SUNDAY, JULY 19, dawned with a perfect Red River sky: high-flying clouds, feathery and rainless. The last of the brief thunderstorms of the previous few days had passed through during the night, bringing just enough rainfall to save the summer wheat without drowning the garden tomatoes. Earthworms dug their way up to sunlight through the moist earth, and excited chickens feasted. The day promised to be cooler, with less humidity, and not the blast-furnace heat more typical of July.

Weather was a topic of conversation when families met at church that morning, weather and the bridge. After services and the midday meal, many families took a Sunday drive to the river to see the bridge, the barricades, and the Texas Rangers, and to enjoy a beautiful day.[1]

People came to the barricaded free bridge every day and night, but this Sunday more than a thousand visitors from Texas and Oklahoma strolled, mingled, and talked there in what seemed like a hopeful holiday gathering. A farmer sold watermelons from an iced tub; people would buy a half or buy a whole, then sit at wooden picnic tables under a big tarp tent to enjoy it. The river was running slow, and some families walked down the embankment to wade at the edges, always keeping a wary eye out for quicksand and snakes. With no motorists on the span, the rangers allowed pedestrians to walk out onto the bridge. Oklahoma and Texas families—some still in Sunday clothes, some in overalls or homemade dresses—howdied one another as they walked the length of the bridge. Children stopped to lean far over the parapets to see the river below until fathers grabbed them by shirt collars or overall straps and pulled them back. The day was bright; the mood was bright.

"Rumors that the bridge has been reopened sweep the city every few hours," wrote reporter Claud Easterly. "Reporters, although certain the report is false, must dash to the river and take a glimpse at the imposing barricade and the rangers serenely standing guard."[2]

Easterly was at the bridge that afternoon, sent by his editor to follow up on (yet another) rumor that the Texas Rangers would be departing. Governor Sterling said the rangers were at the bridge to protect highway department workers, not to keep the bridge closed. The workers were gone, so wouldn't the rangers be leaving, too? Adjutant General Bill Sterling said only that his rangers would stay at the bridge until the governor told them to leave.[3]

The rangers added to the buoyant, carnival-like atmosphere at the bridge that afternoon. Two of the rangers camped at the bridge while two stayed in town, but most afternoons the four of them held target practice with rifles, pistols, and shotguns. Generally, they plugged away with rifles and six-guns at bottles set on a board, or allowed to float down the river, or thrown into the air. Later, they moved the firing range farther down into the river bottom where they practiced machine gunnery.[4]

("They spend the bulk of their time shooting ammunition which, apparently, Texas furnishes in abundance," the editor of the *Durant Weekly* observed.)[5]

Sunday was a hopeful day for the crowds at the bridge. In Austin, they knew, Governor Sterling was assembling all parties for a conference. Surely all of them—the governor, Jake Loy, the bridge people, members of the highway commission, and the attorney general—were working out a solution right that minute. Wherever friends met on Sunday there was speculation as to how long the bridge would remain closed and cries of a cheerful, "It won't be long!"[6]

By early Sunday afternoon, the Austin conference had already fallen apart.

Attorney General Allred, meeting earlier that morning with the highway commissioners and Governor Sterling, declared flatly that the contract with the bridge company was invalid. According to his investigation, the contract had not been signed in an open meeting as required by law. Furthermore, only two of the three highway commissioners had signed the contract. Allred said his office would oppose "any agreement under which the Red River Bridge Company would receive even a nickel from the state." He was ready to fight it out.[7]

"They took this fight into the courts," Allred said, "and now we'll keep it there and obtain opening of the bridge by legal means."

With that, Walter R. Ely, the one commissioner who had declined to sign the bridge contracts the previous year, left the conference room to return to Abilene. The attorney general departed soon after.

An eager Jake Loy and representatives of the Red River Bridge Company arrived that afternoon for the conference, and hope for a settlement dampened considerably when Sterling told them of the attorney general's position. To start the meeting, Cone Johnson, one of two remaining highway commissioners, recited a brief history of events leading to the contracts between Texas and the Red River Bridge Company and to closure of the bridge by injunction. Johnson's statement claimed that Texas had been duped, placing blame for the bridge fracas on double-dealing by the Oklahoma Highway Department and that state's attorney general during contract negotiations.[8]

Without contradicting Johnson's version, Little Joe Bailey described the losses the bridge company was suffering since Governor Murray plowed up the Oklahoma toll bridge approach roads. (Some drivers were getting through by taking back roads to the toll bridge, but Murray's highway men were plugging those leaks as fast as they found them.) Governor Sterling griped that Bailey should be trying to recover those losses from Oklahoma, which set off another round of complaints about Governor Murray.

Jake Loy brought a sheaf of letters, telegrams, and phone messages from constituents in Denison and Sherman showing an increasing anger over the cost and inconvenience of the blocked crossings. They wanted their free bridge opened by whatever means available.

Speaking about the costs of keeping the free bridge closed, Jake Loy asked the governor why Texas Rangers were still guarding the bridge if there was no violence threatened and highway department employees had departed. Sterling snapped back at the state senator. "He said he would keep rangers there and keep the bridge closed until the injunction was disposed of, regardless of what I or anyone else thought of it," Loy later told a newspaper.[9]

The Texas governor was on edge. Attorney General Allred's opinion had cut Sterling's negotiating legs out from under him, but he still wanted a deal. Bailey started the horse trade by telling the governor his reading of the contract had the Red River Bridge Company looking for $150,000 to cover the value of their exclusive bridge franchise and lost revenues. Sterling and Loy thought they *might* be able to get a $50,000 emergency revenue bill through the legislature. Little Joe countered by reminding Loy and Sterling that Judge Kennerly ordered the free bridge closed until Texas honored its contract. Bailey said he *might* be able to talk toll bridge shareholders into accepting $117,000 to settle everything. Impossible, replied Sterling. Given Allred's opinion that the contract was invalid, the most the governor could justify would be $80,000.[10]

And there the matter deadlocked. With no further progress, the bridge company representatives left the building; Sterling left for supper at the governor's residence, and Jake Loy went back to his office in the capitol to compose a telegram.

"I regret that Sunday's meeting called for the purpose of settling the free bridge controversy resulted in no immediate action toward freeing the Denison bridge," Loy wrote to every newspaper in his district. "I had thought for sure that if all parties concerned were present we could settle the question and have the free bridge open not later than Tuesday or Wednesday of this week."[11]

Loy didn't sugarcoat his frustration when he spoke with a reporter that evening. "The trouble is, the governor and several others are worrying altogether too much about Oklahoma and Governor Murray," he said. "We have plenty to do in looking after Texas and need not concern ourselves with Mr. Murray."[12]

News of the failed negotiations traveled faster than the newspapers could publish Loy's telegram; by Sunday evening, word spread through Denison and Sherman.

For all the optimism that morning at the bridge, hopes were crushed when the people of Denison, Sherman, Durant, Colbert, and all the communities on both sides of the river learned that the crossings—free bridge and toll bridge—would remain closed for another round of courtroom hearings. Frustration and anger tinged the hopelessness and helplessness.

In the days since the bridges were closed, a group of community leaders worried that exasperation by the populace might grow into mob action (of the sort that killed the prisoner on trial at the Sherman courthouse just a year before). The breakdown of settlement talks, the blocked river crossings, and further delays opening the free bridge might be just the events to spark some unfortunate violence. A group of community leaders, recognizing the need to give residents an outlet for their frustration, on Sunday evening announced plans for public meetings held to discuss the bridge situation. The first public meeting would be held Monday night in Sherman; another would be held the following evening in Denison.[13]

Jake Loy thought the Oklahoma governor might be helpful in tamping down emotions that were running high along the Red River. He wired Alfalfa Bill again, asking him to "assist in opening" the free bridge by attending the mass meeting in Denison on Tuesday. Murray declined, predicting that the situation would be resolved soon. He correctly saw "an awakening of public opinion" in favor of an early opening on the Denison bridge.[14]

Word of the meetings spread Monday in factories, in cafés, and in grocery markets. Men and women heard of the meeting by telephone, from their postman, and over the back fence. At seven o'clock Monday night, July 20, more than two hundred people milled about outside the temporary courthouse. Judge R. M. Carter—one of the members of the ad hoc organizing committee and the state district court judge who had presided over earlier bridge cases—called the meeting to order. He told the crowd that they should be orderly and polite. Every point of view would be heard, he promised.[15]

Forest Peveto, a Sherman businessman and another member of the organizing committee, was elected chairman. In time, Peveto was able to separate the crowd into three smaller groups, each representing a proposed course of action, and he instructed each group to choose a speaker to represent their views in the form of a resolution that the meeting could adopt.

The smallest group was the "take action tonight" group, ready to march on the bridge barricades with torches and pitchforks. The group speaker received polite applause, but gained little support.

The two other groups were close to equal in number, and County Attorney Ben F. Gafford stepped to the platform first. A decade earlier, Gafford had represented Sherman in the state legislature for two terms before running for the office of county attorney. He was currently the county Democratic Party chairman. Gafford had taught non-lawyer Jake Loy—he referred to him as "our Jake"—the politics of lawmaking and how those laws could be used. Gafford's group proposed that the legislature, then in session, pass an emergency bill allowing the bridge company to sue the state to recover what it was owed, giving bridge owners their day in court while allowing the governor to open the bridge immediately. The plan didn't guarantee immediate action, and the legislative subtlety may have gone over the heads of the audience.

The Honorable Choice B. Randell spoke for the third group. Seventy-four-year-old Randell was a politician of a different era, a six-term US representative, an orator of the old pre-amplified days who had returned to Texas to practice law. Randell spoke of a government responsive to the will of the people. He asked why the communities had not heard from their state legislators, other than Jake Loy. Where were the senators in this matter of interstate commerce? Where were the congressmen? Could it be that Oklahoma's governor cared more about this area than Texas' own governor? Randell implied a distrust of Governor Sterling and questioned Sterling's willingness to lay bare details of the contract, and he proposed that the people demand an emergency committee of Texas lawmakers be appointed to investigate and resolve the issue. Meanwhile, the bridge should be opened right away.

After several hours of debate and discussion, the "Declaration of Resolutions" sent to the governor and lawmakers in Austin and Washington from the Sherman meeting combined both approaches.

"Be assured that no criticism whatever of your conduct in this bridge matter is embodied herein," Judge Gafford added in a personal note to Governor Sterling that accompanied the resolutions. "We had some serious aspersions cast by speakers for the opposition, but we were able to look after the situation."[16]

If the Sherman meeting was as polite and orderly as a garden club meeting, the meeting Tuesday night in Denison was a bar brawl.

The crowd began gathering in Forest Park just south of downtown Denison when the railroad maintenance plant closed at five o'clock. By seven thirty, more than three thousand men and women filled the park and the surrounding streets. (Presumably, the crowd would have been larger if Oklahomans had had a way to cross the river.) The sun had not yet set. The afternoon was hot. People were restless, raucous, and sweating.[17]

B. J. Lindsey, a Denison insurance man and investor, was elected chairman of the meeting. He admitted he was a stockholder in the Red River Bridge Company (drawing boos from the crowd), but said he favored an immediate settlement of the trouble. He believed the bridges should be opened to the taxpayers who built it (drawing cheers). Forest Peveto, chairman of the previous night's meeting, read a telegram from Jake Loy asking attendees what they wanted done. The senator said he and other legislators were ready to act right away.

Peveto also read a telegram from Governor Murray. Alfalfa Bill congratulated and thanked the people of Grayson County for their "loyal cooperation" in his efforts to secure free bridges for the Red River.[18]

The men who had signed up to speak included one who called for the county to secede from Texas and join Oklahoma. Alfalfa Bill will take care of us, the speaker said, where Sterling won't. This proposal, as well as other remarks about the Oklahoma governor, brought wild cheering from the audience, and it lightened the mood of a crowd that might have turned ugly.[19]

Choice Randell repeated his proposal from the night before, and the meeting eventually adopted a resolution much like the one adopted in Sherman the previous evening. "We hereby petition all Texas officials concerned to unite in a supreme effort to give us relief by opening this highway and bridge to the traveling public," the resolution read. "The said highway and bridge must be made available for immediate use, and all obstruction and objection be discontinued at once."[20]

All that was missing from the language of the resolution was the "or else."

The cheers Alfalfa Bill Murray received at the mass meeting in Denison echoed those he was receiving from newspapers on both sides of the Red River.

Initially, editorial writers for the larger newspapers in Texas and Oklahoma harrumphed over Governor Murray's extralegal showboating. The *Daily Oklahoman*—no friend of the Murray administration—recognized the popularity of his preemptive bridge opening: "All the facts and all the circumstances combine to make the gubernatorial action popular among those who care more for results than they do for methods." "What Governor Murray has done is striking," opined the *Dallas Morning News*, "but it is not law and it is questionable whether it is justice."[21]

As days passed and small-town publishers added their voices, it was apparent that Alfalfa Bill was winning the war of public opinion. One of southern Oklahoma's largest newspapers, the *Daily Ardmoreite*, equated Murray's opening of the Denison bridge with Teddy Roosevelt's charge up San Juan Hill. "The people of Oklahoma and Texas are with Murray almost to a man," it added. The *McIntosh County Democrat* swooned over Murray's actions: "Oklahoma's governor has again risen to champion a worthy cause and has again proven that he is one of the greatest students of government in America."[22]

For many, Murray's actions marked him as a politician willing to take unorthodox action during exceptional times. Letters to newspaper editors and letters to Murray also applauded those actions. Herman Martens, manager of a Dallas printing ink company, said when Murray opened the bridge, "we Texans thought he was headed for the jailhouse, but we found out he knew what he was doing and we learned he has plenty of nerve." West Texan

Z. W. Davis thanked "the Lord or whoever is responsible that we have one official in the country who takes some interest in the people's lives. I am referring to Governor Murray of Oklahoma."[23]

And the Oklahoma governor wasn't through with the toll bridge matter. He intended to seal the river crossing off completely.

Murray had ordered the Oklahoma roads approaching the toll bridge to be made impassable. Bulldozers churned up two hundred feet of asphalt road, but there were still a few motorists willing to drive around the obstruction. Murray told highway workers to build a fence on both sides of the torn-up road. Locals had discovered a route around the approach road, a four-mile detour that required crossing a pre-automobile one-lane wooden suspension bridge across Sandy Creek. Alfalfa Bill was probably doing local motorists a favor by having a highway engineer inspect the swinging bridge, condemn it as unsafe, then oversee the dismantling. The bridge at Carpenters Bluff, five miles downriver, was open to determined motorists, but to reach the bridge meant driving down miles of deeply rutted dirt and tire-spinning sand.

Reporters seemed to love the homespun Oklahoma governor with the walrus mustache, and Alfalfa Bill wasn't above doing a little headline baiting. Still chafing under Ross Sterling's assertion that "you have extended your jurisdiction beyond all reason," Murray mused to a reporter that maybe he'd send the National Guard down to the free bridge and remove those barriers again. But what about the Texas Rangers? the reporter asked.

"I'll arrest 'em," Murray said. "Those Texas Rangers are actually in Oklahoma." That sent Murray off on a thirty-minute history lesson on Thomas Jefferson, the Louisiana Purchase, the Rio Rojo, and the Adams-Onís Treaty of 1819. Asserting Oklahoma had jurisdiction over the south bank of the Red River, Alfalfa Bill said, "If they put anything else in the way of an obstruction of the bridge I will remove it in the face of Texas Rangers and any other military authority of the state."[24]

As expected, the reporter took his comments seriously and moved the story on the wire service. Murray's words earned front-page headlines in newspapers across the country.

Each new headline brought more supportive letters, telegrams, and reporters to his office, and Murray was glad to give them another story. After the Austin negotiations collapsed, a reporter asked Murray whether he had any suggestions about how to end the stalemate. An excellent solution, Murray said, would be to hold a quilting bee, where women from Oklahoma and women from Texas would meet on the new free bridge, "removing anything that prevents their getting together."

"The National Guard wouldn't shoot a woman if she came along and took a few barricades out of the way," Murray told the smiling reporters. "And neither would a Texas Ranger. After the barriers once were down, the traffic could cross and the toll bridge would be whipped right there."[25]

Ross Sterling was not amused, and he answered Murray's jocular proposal with a grumpy letter. "I do not believe the good women of the two sister states will permit themselves to be incited by any such tomfoolery," Sterling wrote. "I trust they will not be led astray by such a questionable proposal."[26]

Despite the humor Alfalfa Bill (and newspaper headline writers) found in the situation, the Durant-to-Denison bridge closure was a serious affair, threatening hundreds of Oklahoma farmers along the Red River. But Bill was also dealing with issues that could affect the entire state.

The Henryetta food riot was a fire bell, a signal warning of a populace tired of legislative dawdling and frustrated with hollow political promises. It had been a one-day affair and there had been no violence, but those jobless men weren't going anywhere, and their families couldn't be allowed to starve.

Murray phoned the Henryetta mayor on Wednesday to confirm action on the plans he and the mayor had discussed over the weekend. At the governor's suggestion, the mayor had formed a relief committee to solicit donations and distribute aid. Every jobless family would be given enough aid to provide the necessities of life. Murray wanted to be sure the committee enforced his one ironclad rule: "No work, no eat." (The governor was adamant that no state money would go to layabouts.)

The mayor had compiled a list of work to be done around the town: cleaning streets, raking alleys, dismantling an old building, and clearing vacant lots to plant as vegetable gardens. The major project was the building of a road from town to the municipal lake three miles away, and workers would be paid a dollar a day. Governor Murray promised to divert highway funds by executive order to pay for labor and road-building materials.[27]

The same day—and also using an executive order—Murray declared a moratorium on rental payments for 340,000 acres of school land owned by the state. The land had been retained by Oklahoma at the time of statehood and most of it was leased to small farmers, many of whom were facing eviction for non-payment of rent. The governor's moratorium would give more than two thousand farm families some breathing room before facing eviction, and the farmers would be charged a modest interest rate for the deferred payments.[28]

Farmers were often quiet about their problems, but murmurs of hard times and calls for systemic changes echoed loudest in the cities. On Wednesday, July 22, a band of fifty jobless men marched through the streets of Oklahoma City to a wholesale grocery company's warehouse to ask for food. Five men went into the building while the rest remained outside with baskets to carry the food.

They were members of the Unemployed Council, a group of jobless men formed after the hunger march in Oklahoma City the previous winter to aid in obtaining relief for destitute families. Most of the men had found work during the spring, but those jobs had run out. Their families were in desperate need of food, they said.[29]

Acting on a tip, police were waiting inside the warehouse. They arrested and beat the five leaders and ordered the rest to disperse immediately. That afternoon the police chief issued a bulletin ordering officers to immediately arrest any gathering of the unemployed or those encouraging attempts to take food from stores.

When Governor Murray heard about the arrests he sent an assistant state attorney general to the jail to free the prisoners. By executive order he rescinded the chief's order and granted an executive pardon to the five arrestees. Murray promised immediate aid, but he was running out of places to find it.[30]

"In a land swept by drought, gnawed by grasshoppers, devoured by army worms and crushed by business languor, Governor Murray is as a cool brawling stream," a New York columnist wrote about Alfalfa Bill. In fact, the governor was more of a magician, showing the bright, shiny object in his right hand to distract from what was happening with his left. It was better for the press to focus on events at the Red River bridges than on hungry mobs on the streets of Oklahoma towns.[31]

Closure of the Red River crossing was becoming an expensive proposition for businesses in Texas and Oklahoma. The federal highway between Durant and Denison, US Hwy. 75, was a major north-south route for passenger cars, commercial carriers, and interstate buses. The free bridge was barricaded by Texas Rangers, Oklahoma had cut all access to the toll bridge, and all traffic from one state to the other was forced to detour twenty miles to the west over poor roads to cross a one-lane wooden bridge at Preston. It was summer, prime tourist season, and business at the filling stations, diners, and tourist camps along the highway was bleak.

The president of the Denison Chamber of Commerce wired Governor Sterling: "Large part of Denison trade comes from Southern Oklahoma and this trade is being jeopardized at present and perhaps permanently. Situation is tense. Traffic paralyzed." The Retail Merchants Protective Association of Sherman begged both governors, saying, "Anything you can do to expedite opening of the Red River bridge at Denison will be greatly appreciated."[32]

H. A. Randall, a traveling salesman from Durant, made it a point to visit the editor of every newspaper in his territory—from Wichita, Kansas, to Dallas—to discuss "the abominable situation caused by blockading the free bridge" on US Hwy. 75. "No consideration has been shown the traveling public or taxpayers in all these negotiations," he said. He termed the barriers on the Red River bridge "an absolute injustice."[33]

The state of Oklahoma added up its losses from the bridge closures. According to Carl Giles, committee chairman of the Oklahoma Chamber of Commerce, the average auto tourist will spend $7 a day and not less than a day and a night in the state. Eager to make up time lost to detours, north-south travelers were spending less time in Oklahoma. Some tourists, aware of national headlines about the bridge controversy but unsure of exactly which bridge was closed to travel, chose to avoid the state altogether. The Oklahoma chamber began placing ads in Dallas, Kansas City, and Chicago newspapers with detour instructions.[34]

The Dallas Chamber of Commerce was feeling the loss, too. "Dallas is almost as much a victim of the controversy as Denison and Sherman," one official said. Denison was the Gate City, and traffic entering Texas there would continue south on US Hwy. 75 to Dallas. Tourist traffic to Texas was now being routed to the west, to Fort Worth. The Highway 75 Caravan was expected to arrive in Dallas Friday night and continue northward toward Winnipeg on Saturday. Organizers said the caravan had grown to 150 autos, and they were phoning daily about the bridge closure. Ben Critz, the chamber's general manager and the caravan's event host in Dallas, was on his way to Austin to discuss the bridge closure with highway commission members and Governor Sterling. He planned to point out that Dallas was materially affected by the failure of officials to reach an agreement.[35]

There was one bright spot, however. Oklahoma, with lax age restrictions on marriage licenses and no waiting requirement, was a favored destination for sudden elopements or a quick "I do" by teenage lovers from North Texas.

Durant had held the title of Marriage Capital of Oklahoma, but the bridge closings made Madill, just ten miles north of the Preston bridge, the primary destination for courthouse-bound couples. "We have just as good a license as Durant, and we have preachers with just as nifty ceremonies," said J. M. Scott, a Madill town booster.[36]

An editor of the *Dallas Morning News*, tongue firmly in cheek, worried about what might happen if the Preston bridge were to be closed: "If Alfalfa Bill Murray keeps on, and the Texas Rangers keep on, our gin-marrying Texans are going to have to detour to Arkansas."[37]

Meanwhile, the hotels, restaurants, clothing stores, and florists of Madill were enjoying a temporary bonanza.

Ross Sterling could have used a bit of good news. By Tuesday evening—with Texas Rangers blocking the bridge, failure to negotiate a settlement with the toll bridge owners, Allred declaring the contracts invalid, Governor Murray's chaff, and Senator Loy baying at his heels like a bloodhound after an escaped convict—the governor could find little to cheer him. Sterling was hearing from friends, supporters, and voters from all over the state with opinions, usually negative.

As vice president of one of the largest road construction companies in the Southwest, R. S. Womack of Denison had become acquainted with Ross Sterling through the Texas Highway Commission. Womack was an early supporter of Sterling's gubernatorial campaign, and the two men could speak businessman to businessman. Womack's telegram to Sterling cut like a saw blade. "Very much disturbed over failure to open bridge," he wrote.[38]

J. H. Durning, manager of Texas Power & Light in Sherman, had hosted a dinner for Sterling during his campaign. Now he was writing on behalf of the businesses there. "Traffic being forced to unreasonable detours and situation growing more acute hourly," Durning told the governor. "We urge immediate action toward settlement."[39]

Sterling regretted the voters he was losing. "I wish to enter my protest on the action you have taken in the Oklahoma bridge matter," wrote Leo C. Mills of Dallas. "I am sorry to say I will never cast another vote for Ross Sterling if you run for governor or anything else as long as I live."[40]

J. C. Jones was, like Sterling, a loyal Texan. "I voted for you and had hopes to stand by you," he wrote the governor. "But I think this [bridge issue] is a sore eye for Texas. I hope you see this as thousands of other good

Texas people have and try and keep Texas' good name as we have always had it."[41]

Editorials in Texas newspapers weren't quite so negative; most didn't mention Sterling by name. Instead, writers were excoriating the Texas Highway Commission—of which the governor had been chairman at the time—for not paying what they owed or for getting into the mess in the first place. One editor's opinion was that "the members of that commission ought to resign their places and Governor Ross Sterling ought to permit Oklahoma's governor to appoint men to succeed them."[42]

Published letters to the editor were just as unwelcome. "Oklahoma is to be congratulated on having a real man for governor," a Dallas reader wrote. "It's too bad there are not more governors like him."[43]

Sterling was tired of hearing about Governor Murray. Sterling snapped back to at least one correspondent. "I believe you were wrong from every angle in dispatching State Rangers to the Denison Red River bridge," wrote W. C. Tatum, a respected medical doctor from Fort Worth. "It is probable that the people of the two states will consider the Governor of Oklahoma much nearer right in his acts and attitude." Most letters received a standard response, but Sterling wanted to answer Dr. Tatum personally. The governor picked up a pencil and scrawled his response at the bottom of Tatum's letter, breaking the pencil lead at least once and digging deeply into the paper sheet. "You would change your mind were you conversant with all the facts concerning the bridge controversy," Sterling wrote without any preamble. "The good people of Texas want law and order, and it will be a sad day for our state when we disrespect our courts. I'm sorry you feel my actions were wrong."[44]

Nine months after his election there was little remaining of the ebullient "Fat Boy from Houston who'll work for you." Whatever well of good will Sterling was drawing from was just about dry. He had won the governor's office in no small part because of the competent, honest way in which he ran the Texas Highway Department. But his highway department was being blamed for depriving the people and businesses of Texas the use of a $200,000 bridge that they had already paid for. Citizens and press doted on Governor Murray for his willingness to take decisive action and to solve problems using executive orders. The times and circumstances were calling for a man of action, but the plodding legal process hobbled Sterling. He was a law-and-order man; it might take a little longer, he believed, but the law was to be respected and followed. Meanwhile, the bridge north of Denison had been blockaded for three weeks and guarded by Texas Rangers for six days. Traffic was cut off between the two states. Sterling was still stuck with his tar baby, and there seemed to be no way to come unstuck.

Until state senator Jake Loy walked into the governor's office.

Loy had spoken with Judge B. F. Gafford, the sixty-year-old county attorney and former state representative from Sherman, and Gafford had a suggestion for breaking the deadlock. Gafford had attempted to explain the solution to Loy earlier in the week, but Loy, who was not an attorney, didn't fully understand on first hearing. Finally, by Tuesday evening, July 21, Loy could relate Gafford's proposal to Governor Sterling.

Judge Kennerly's injunction forcing closure of the free bridge had been granted to the Red River Bridge Company on the argument that it had no remedy at law to enforce provisions of the disputed contract. The toll bridge company wanted its day in court, a chance to prove damages and finally collect for its losses of property and revenue. The state of Texas and its agencies were protected by sovereign immunity; they could not be sued by the toll bridge company. Therefore, there would be no day in court. If, however, the state legislature passed an emergency bill to waive that immunity for a special case, Judge Kennerly could dissolve the injunction, opening the free bridge and ending the standoff. The toll bridge would get its day in court.

Gafford had spoken to several of the toll bridge officers and shareholders, he said. They assured him they would ask to dissolve the injunction if the legislature waived immunity. Loy's job was to bulldog the emergency bill to waive immunity through the legislature as soon as possible, and he had spent the day consulting with legislative leaders. If Attorney General Allred would sign off on the plan, and if Governor Sterling would twist a few arms, Loy thought he could get the bill to the floor by Friday and the bridge open by the following Monday.[45]

This was a lot of "ifs," but it was Sterling's first good news in a while. He told Loy he would arrange a meeting the next morning.

At least one Texas Ranger patrolled the free bridge barricade between midnight and sunup. Occasionally, a hobo might try to cross the bridge on foot and the ranger would turn him back. Sometimes a motorist, thinking the bridge would not be guarded at that time of night, would try to sneak across. The ranger would turn him back.

Hours could pass in the warm and sticky dark, and the ranger on patrol might hear nothing but the sounds of the river. At other times, he might hear the distinctive engine sound of a Model A Ford a mile downstream, a bootlegger crossing the riverbed from sandbar to sandbar, smuggling whiskey from Oklahoma into Texas.

On July 21, three young men fished the river all night and, from above on the bridge, the ranger probably heard wisps of conversation, slaps at the ever-present mosquitoes, and the splash of hooked fish.

The three fishermen were from Texas. They had walked across a shallow part of the riverbed before dark to fish on the north bank of the river. Before dawn, they gathered their gear to return home. It would have been fairly easy to walk up the bank and cross the river on the bridge, but the ranger would have turned them back. Instead, the fishermen chose to make their way back across the riverbed in the dark.

Twenty-six-year-old Olan Carter, a farmer from the White Rock community, was about two hundred feet ahead of his companions. Picking his way across the sandbars, he tripped and fell into a deep pool of swirling water. His companions tried, but failed, to locate him in the dark. Carter's body was found two days later downstream near Carpenters Bluff.

Olan Carter was the second casualty of the Red River Bridge War.[46]

CHAPTER 10

Ready to Go Killing

GOV. ROSS STERLING entered his office on Wednesday, July 22, with the hope that he might escape his personal bridge tar baby after all.

Sterling and Jake Loy convened a conference that morning with Assistant Attorney General Sidney Benbow; state representative Elbert Barron, whose district included Denison; and Cone Johnson, chairman of the highway commission. Sterling asked Jake Loy to describe his legislative plan for ending the bridge standoff.[1]

Loy outlined the plan, then looked around the table for comment. Benbow was skeptical, saying his boss, Attorney General Allred, wouldn't be too enthusiastic about the scheme, believing he could launch a successful appeal of Kennerly's injunction or, in partnership with Oklahoma, persuade a US Supreme Court justice to lift the injunction until the full court could hear the matter. Either course might take several weeks, however, and there was no guarantee of success. A bill by the Texas legislature allowing the state to be sued would, if written correctly, be legal, Benbow said, but would it get the injunction lifted immediately? Loy had it on good authority that representatives of the toll bridge company had promised to withdraw the injunction suit in the event a law should be enacted allowing them to sue the state of Texas.[2]

Loy and Barron said they had spoken with a large number of legislators who promised their support. Lawmakers were hearing the same expressions of frustration and anger they were hearing from constituents. Another mass meeting was scheduled for Monday, and there was a concern that the meeting might encourage some to take the law into their own hands. Loy predicted no difficulty in passing an emergency bill in just a few days.[3]

Early that afternoon, Sterling asked Benbow, Loy, and Cone Johnson to draft the legislation while the governor dictated a message to accompany it. They would submit the bill to the senate that day.

Drafting and reviewing the bill by all parties took longer than expected. Loy rushed the final version to the senate chamber late that afternoon, but arrived five minutes after adjournment. The bill would have to wait until Thursday to begin winding its way through both houses of the legislature. That afternoon and into the evening Loy skittered between legislative offices, senate chambers, the governor's mansion, and the restaurants and hotel lobbies along Congress Avenue, hoping to grease the skids for legislation ending the Red River bridge imbroglio.[4]

That same evening, in Oklahoma City, Governor Murray heard a disquieting rumor. The Red River Bridge Company planned to petition the federal district court in Muskogee for a restraining order. The bridge owners intended to require Murray and the Oklahoma Highway Department to remove all barriers to their toll bridge and repair all damage done to the approach roads.

Newsmen had heard the same rumors, and they asked Murray whether he might declare martial law. "Let's don't talk about it now," an uncharacteristically tight-lipped Murray responded.[5]

"President Hoover carried out his prosperity promise," US Senator Thomas P. Gore of Oklahoma told the joint session of the Texas legislature. "He carried it out on a stretcher."[6]

In an unfortunate coincidence, the popular (and entertaining) Democratic senator from Oklahoma had been invited to address Texas legislators about a national need to help oil and cotton overproduction on the same day that Jake Loy had hoped to speed his Red River bridge bill through both houses of the legislature. Gore was taking up valuable floor time that Loy dearly needed.

"Hoover also promised to put the American people back on their feet," Gore continued. "He has, to the tune of seven million unemployed walking the streets in search of jobs."[7]

Jake Loy's morning had gone according to plan, thanks to some help from the president of the state senate. The senate was gaveled to order at nine-thirty, and it took less than five minutes for members to waive reading of the governor's message, accept the committee report, suspend the rules calling for reading the legislation over three different days, and vote on SB 9, "An Act Permitting Red River Bridge Company and/or J.R. Handy, Receiver, to Sue State Highway Commission and State of Texas."[8]

"All this occurred while most members were reading the morning papers," according to one observer, "and many did not look up except to answer 'aye.'"[9]

The bill passed the state senate on a vote of twenty-eight to one, and was referred to the House of Representatives just in time for the state senate to adjourn at ten o'clock. Loy walked with the other senators across the rotunda lobby to the house chamber, where Senator Gore would be speaking.

In the house chamber, members sat at their desks while young pages placed chairs in the aisles for the senators who had come to hear Gore. Jake Loy, aware of every minute slipping away, paced at the back of the chamber.

On his way to a Houston bar conference that weekend, Gore said he wanted to let lawmakers "in on the lowdown" of the Sterling-Murray bridge controversy.

"Governor Sterling is trying to keep the Red River bridge closed so that broke ex-Texans can't come back from Oklahoma to live off the wife's kinfolks," Gore told the audience. "And Governor Murray is determined to keep the bridge open so Democrats can migrate from this Republican state and help increase Oklahoma's Democratic majority."[10]

Gore kept the audience of lawmakers laughing while he laid out an indictment of Republican tariffs and "Hoover prosperity." Meanwhile, Jake Loy fumed, worried that he'd be unable to get his bill through the house before the weekend. Given the mood of the people along the Red River, violence was a real possibility if the bridge was not opened soon.

Eventually Gore's speech ended, senators left the chamber, and Loy made his way to the desks of Representatives Olan Van Zandt and Elbert Barron. Loy had chosen Barron, who represented Denison, to introduce in the house a bill that was identical to the one passed by the senate. Both bills would be returned to the senate for final passage, then sent to the governor for his signature. "The bill has now gone to the house for action there. I hope to have the bridge open in the near future," Loy said.[11]

Van Zandt, who represented the area around Sherman, didn't seem too happy about not being named as a co-author on the bill. Senator Loy had been quoted and headlined in the Sherman and Denison newspapers every day, as if he alone was responsible for settling the bridge standoff. Van Zandt knew that those news stories would prove valuable come reelection time, and tried to claim some of the credit for himself. "Being reliably informed that Senator Loy would not sponsor legislation, we prepared a bill for that purpose," Van Zandt told a reporter for the *Sherman Democrat.* "Then, in conjunction with the governor, we submitted the bill to the attorney general's department."[12]

Shortly before noon, as the House of Representatives went back into session, Van Zandt informed Loy and Barron that he intended to sponsor an amendment to the senate bill, a move that would link his name with the final legislation. Loy exploded. "Do not tie this thing up!" he said. "It is a deplorable situation up there. People are about ready to go killing each other."[13]

Barron introduced the bill in the early afternoon, and the skid-greasing he and Loy had done over the previous thirty-six hours paid off. As in the senate, the house quickly waived the mandatory three readings and was ready to vote. It was then that Van Zandt announced his amendment to the bill. The amendment was a minor change in wording, but by the time Van Zandt finished reading it, another house member wanted to make a minor change. Then another member with still another change.[14]

When the house adjourned for the day, the final bill had not been sent to the senate. Loy would try to finish the job the next morning. In a conversation with reporters, Van Zandt denied he was responsible for failing to pass the bill. "If anyone was to blame," he said, "it was Senator Loy."[15]

The senator wasn't too concerned about what Van Zandt said; he was exhausted. Loy had never considered the amount of physical labor necessary to be a state senator until the previous week, working nonstop for five grueling days dragging Governor Sterling's fat out of the bridge bonfire while acting on his constituents' need to open that silvery new bridge over the Red River. The good news should reduce the heat under any conflict that might be percolating among residents along the river. As Loy prepared to leave the capitol, he was expecting a quiet evening in his hotel room that night and easy passage of his bill in both chambers Friday morning.

Jake Loy's evening wouldn't be as uneventful as he hoped. A reporter from the *Austin Statesman* caught him at his office door to ask what the senator thought about Governor Murray's proclamation of martial law around the Red River bridges. For a moment, Loy was speechless. Then he turned around and returned to his office.

Reporter Claud Easterly was in the newsroom, reading through the afternoon edition of the *Denison Herald,* when the teletype bell dinged and came to life. Easterly joined others gathered around the clackety machine as news printed out line by line.

The Associated Press in Oklahoma City was reporting that Governor Murray was declaring martial law along a strip of land on the north side of the river. Armed Oklahoma National Guardsmen, under the command of Adjutant General Charles F. Barrett, would arrive at the bridges that evening or early the next morning. Murray ordered Barrett to exert "military control against all interference whatsoever, except only an order otherwise by the superior executive power of the President of the United States."[16]

Easterly grabbed his hat, checked his pocket for a notebook, and rushed off to the bridge.

Most people considered it would be just a matter of days until the bridge legislation was passed and the Houston judge allowed the free bridge to open. Murray, however, knew of pending federal court action in Muskogee that could delay the opening of the free bridge and reopen the toll bridge. He was calling up his army.

"It is threatened by the bridge owners by some process, court order or otherwise," Murray wrote in the preamble to his martial law proclamation, "to open the obstruction placed upon the highway approaching the toll bridge, which approach has been ordered closed by executive order of the State of Oklahoma." Such action, he continued, "would be an invasion of the rights of the sovereign State of Oklahoma."[17]

Murray wasn't about to let some federal judge or private company force him to reopen the toll bridge. "It is to the best interests of the people of the state of Oklahoma to secure access to the public property aforementioned and to prevent the exaction of tolls from her citizens by the Red River Bridge Company," the order said. The proclamation was signed by Murray as "Commander-in-Chief of the Military Forces of Oklahoma."

The governor ordered the adjutant general to "summon sufficient forces" to block approach roads as necessary to assure that the toll bridge remained closed to traffic.

"Sufficient forces" waited for General Barrett in Durant, home to three companies of the state guard. As Barrett drove from Oklahoma City that evening, Durant-area guardsmen were being contacted and told to muster at the Bryan County courthouse. The companies being mobilized included an infantry company, a howitzer company, a medical unit, and a headquarters unit, 290 men in all. More than half were veterans of the Great War, and had received extensive arms training.[18]

Claud Easterly was the first reporter to arrive at the Texas Ranger encampment on the south end of the free bridge that evening.

No, Captain Tom Hickman said, he had heard nothing regarding the declaration of martial law. Neither had anyone informed him of a change in the present force of rangers. His orders continued to be to keep the barricade on the free bridge intact. He assured the reporter that the "four Texas Rangers would let no one pass."[19]

Texas Adjutant General William Sterling of Texas, attending a meeting in San Angelo, left immediately for Austin when a wire service reporter informed him of the martial law declaration. He refused to make any comment except to say that he expected to confer with Governor Sterling. The adjutant general was probably unaware that the governor had departed Austin; he was traveling to Houston to speak at the same bar conference Senator Gore was attending.[20]

Regardless of how the situation developed, there would be no reinforcements for the Texas Rangers at the bridge that night.

W. G. Wood, president of the Hwy. 75 Association, placed frantic phone calls to the Sherman and Denison chambers of commerce that evening, seeking up-to-the-minute bridge news. The Galveston-to-Winnipeg motorcade had departed Galveston, and would arrive in Dallas Friday evening. Would the bridges be open by Saturday morning? Or would the Hwy. 75 Association have to admit to detouring fifty miles off the very highway that the association was promoting?[21]

Lute Loy, chairman of the chamber's traffic committee (and brother of Senator Jake Loy), told Wood that he was hopeful that the bridge would be open in time for the motorcade. The Speaker of the Texas House of Representatives was conferring with Governor Sterling, he said, about a temporary waiver allowing the motorcade to pass over the free bridge.

Loy didn't tell Wood that the motorcade might be driving into the middle of a war zone.

News that Oklahoma troops were heading for the bridges spread quickly through towns and communities on both sides of the river. A larger-than-usual crowd drove to the bridge for their evening stroll on the span and to watch whatever fireworks might develop. The first arrivals saw two of the rangers, assisted by two sheriff's deputies, re-installing the floodlights they removed earlier in the week. The remaining rangers announced that no one would be allowed on the bridge that evening. No pedestrians, no autos, no one.[22]

When the acetylene flood lamps were connected, the deputies moved the crowd two hundred yards back up the approach road and formed a barrier using their own automobiles. The rangers settled in by their barrier to wait for sundown.

Claud Easterly noted that for the first time since they had taken their post, all the rangers were wearing holstered sidearms.

General Barrett arrived in Durant well after dark on Thursday evening to meet with the guardsmen who had mustered on the lawn of the county courthouse. Lt. Col. John A. MacDonald (a Durant attorney and state senator) handed Barrett a phone message from Governor Murray to order out only thirty to forty men instead of the entire contingent. (Murray was worried about costs.)[23]

Together, Barrett and MacDonald chose thirty-six men from the infantry and artillery companies. The rest were sent home with Barrett's thanks. The troop consisted of three officers and thirty-three enlisted men, and included two full-blooded Choctaw. The men wore full uniform and bore regulation rifles and pistols. A rolling field artillery piece was chained to the back of a truck and readied for departure to the front while an adjutant directed enlisted men to load the truck with extra ammunition, tents, cookstoves, and several trench mortars brought in from the armory.

Barrett's arrival in Durant was preceded by the arrival of reporters from Ardmore, Tulsa, and Oklahoma City, and Colonel MacDonald issued military press passes to each of them. Barrett told the newspapermen that he would speak to them the next morning after his troops were deployed. Barrett expected to return to Oklahoma City soon, and command would be turned over to Colonel MacDonald.

MacDonald declined to comment, saying only it was a "rather tense situation." He signaled his officers to begin loading the men onto war surplus trucks for their trip to the front line twenty miles away.[24]

Alfalfa Bill's army was on the move, heading for the bridges over the Red River.

The Texas Rangers knew only that an army was heading their way. As night wore on, they remained on alert, standing behind the barricade and listening for what might be happening on the north bank. There were the usual night sounds—insects buzzing around the floodlights, the hiss of river over sand, and the occasional automobile noise far in the distance—but nothing that would tell them an armed force might be coming their way.

Texas Rangers carried a reputation as a shoot-first-ask-later outfit, and Claud Easterly—who was also spending the night at the bridge—expected to hear the sound of gunfire at any moment. Near midnight, the small group of Texans heard angry shouts coming from the Oklahoma end of the bridge. It was a man's voice, yelling at the rangers and coming closer. Flood lamps illuminated the underside of the bridge, but the roadway was murky in shadow. Eventually, the rangers could make out a small group walking toward the Texas side.[25]

The rangers drew their weapons.

The makeup of the group was unclear and their words indistinct until the group reached the midpoint of the bridge.

"Set your machine guns, men; we're not afraid of Texas Rangers," a belligerent voice called out.

The man who walked out of the shadows was not wearing a uniform. He was in his forties, dressed in rough work pants and denim shirt with sleeves rolled up. He had the thick arms and ruddy complexion of an outdoor laborer. He appeared drunk, staggering as he continued hurling abusive language at the rangers.

Three other people were with the man, all attempting to hold him back and doing an unsteady job of it. The two women wore dresses and hats, and they were made up to look like downtown girls. The other man wore cotton pants, a dress shirt without a collar or tie, and a straw Sennit perched on the back of his head. The three were trying to turn the fourth around, urging him to stop cursing the rangers. The scuffling group staggered to within thirty feet of the armed Texans.

The rangers holstered their sidearms as Tom Hickman stepped around the barrier and approached the swearing man. "Turn around and return to Oklahoma," Hickman told the group, earning another barrage of loud and abusive language.

Hickman stood face to face with the cursing man, asking that he rein in his language and reminding him that women were nearby. This time the man directed himself to Hickman alone, calling into question the captain's manhood and ancestry.

Hickman grinned for a moment; then his arm flashed upward and struck the man's face. (The Texas Ranger captain later described the blow as an open-handed slap, and the reporters present confirmed his description. But his "slap" had the effect of dropping the drunk to the roadbed and bloodying his nose.)

The man in the straw hat helped Captain Hickman lift the downed man and walk him back to the Oklahoma side. The man was from Durant and worked at the rail yard in Denison. The closed bridges had kept him from work for almost a month, and he wasn't sure whether he still had a job. The two men had met the two girls at a roadhouse earlier that evening, and things whirled out of control from there. Parking the foursome on the Oklahoma end and warning them to stay off the bridge, Hickman returned to the barrier.

The unnamed Durant man with the bloody nose was the third casualty of the Red River Bridge War.

For the rest of the night, as Thursday turned into Friday, the rangers heard nothing else across the river but the muffled sounds of men moving over rough ground and the grinding of the gears of an overloaded truck.

CHAPTER 11

Hold the Fort

BY SUNUP Friday morning, Alfalfa Bill's National Guardsmen were spread along the martial law zone, a strip of Oklahoma land two hundred yards wide and more than a mile and a half long that covered the road leading south to the toll bridge entrance. It was, in all respects, a peaceful deployment; the martial law zone came no closer than a half mile from the rangers guarding the free bridge barricade.

"Accoutered as for war, Oklahoma soldiery today held a thin gray line of roadway against a phantom foe," a wire service reporter wrote.[1]

General Barrett arrived at the restricted zone to inspect his Oklahoma soldiery early that morning (after a good night's sleep and a hot breakfast in a Durant hotel). "The troops are instructed to let no one enter," he told reporters. "Persons living in the zone who must take the road will be allowed to pass, but no one will cross the toll bridge."

A group of reporters—they called themselves "war correspondents"—accompanied Barrett along the line as he inspected the troops. Each soldier carried a red flag to warn off anyone who approached, and all were armed with regulation rifles and pistols while on patrol. A reporter asked whether the rifles were loaded.

"Try and get through without a pass and find out," a serious General Barrett responded.[2]

One of the guardsmen later admitted that none of the firearms was in working order. They had been stored away so long the oil had turned to tar. "We were pretty sure we wouldn't have to fight," he said.[3]

Colonel MacDonald had appropriated a small brick building, a maintenance shed overlooking the entrance to the toll bridge, to serve as headquarters. As Barrett ended his inspection, a reporter asked MacDonald whether he was expecting any action at the bridge. MacDonald deferred to his superior officer.

"I believe it will be a dull campaign," the adjutant general answered. He added that he thought it would be all over in a few days, as soon as the Texas legislature passed its legislation allowing the bridge company to sue the state. "If the injunction for the free bridge is dissolved, the need for the guard will have passed."[4]

Barrett's instructions to MacDonald came directly from Governor Murray: "Hold the fort but keep the cost down."[5]

No one in Austin was more committed to getting the bridge legislation passed than state senator Jake Loy. A series of minor amendments in the state house the previous afternoon had prevented passage of the bill there. He and Barron would have to get a favorable vote in the house and walk the bill back to the state senate, where the different versions could be reconciled before the senate voted for passage. Loy was working against the odds; the senate was due to adjourn early that day for a three-day weekend.

Another factor working against prompt passage of the bill was a surprise telegram from Joe Bailey Jr. In a wire that arrived in the capitol sometime after one o'clock that morning, the Red River Bridge Company attorney told Governor Sterling he had only just learned of pending legislation allowing Texas to settle the contract and end all toll bridge litigation. Bailey had read press reports that his client had agreed to the plan.

"No such agreement has been made," Bailey wrote, "and the press reports are greatly in error." His thousand-word telegram laid out the chronology and terms of the contracts signed by Sterling and the highway commission. Bailey called out Governor Sterling personally, saying the bridge company had entered into the contract because it had been assured "you had a lifelong reputation for carrying out every contract to which your name was affixed." Bailey demanded that he and his client "be accorded a hearing before the passage and approval by you of any such act, that we may present our side of the matter before such action."[6]

It was the final paragraph of Bailey's telegram that fell like an anvil on Jake Loy.

"We have never given any statement to the press in this connection," Bailey wrote, "but in view of the wide publicity by other parties in an attempt to put the bridge company in an unfavorable light, we are taking the liberty of handing to the press a copy of this telegram."

Full terms of the contracts with Red River Bridge Company had never been made public, and the disclosure would certainly blacken Governor Sterling's "life-long reputation" for honoring contracts. It might also call into question other of Sterling's dealings as highway commissioner, expedient actions that, although not quite illegal, might lead to still further questions about his integrity as governor.

Loy's reputation was in danger, too. Judge Gafford and other Denison civic leaders had assured him that the bridge ownership would accept the opportunity to sue the state of Texas. Loy promised his fellow legislators and Governor Sterling of the same. Now it might appear that Loy had misled them.

There was the possibility, also, that some headline-grabbing representative or senator might accede to Little Joe's demand to take his case to a committee before final passage of the waiver legislation. That could delay the bill—and the opening of the much-needed free bridge—indefinitely.[7]

In one way, however, Bailey had miscalculated. His telegram to Sterling—and the copies sent to state newspapers—had been sent in the early hours of Friday morning, too late for morning newspapers to publish the story. If Loy could get the bill passed by both houses of the legislature before the afternoon papers went to press, things might work themselves out in the end.

Loy was waiting at the doors of the house chamber before the usher called the body to order at nine o'clock.[8]

As it turned out, news reports of armed Oklahoma National Guardsmen arrayed across the banks of the Red River and facing south was enough to focus the attention of the house and senate. The Texas House of Representatives moved the amended bridge bill to the top of the agenda, passed it unanimously, and returned it to the senate. The senate accepted the amended bill without discussion, passed it (with one dissention), and engrossed it before the afternoon newspapers were circulated.[9]

Loy's bill had been passed, but it wouldn't become law without Sterling's signature. The governor was away from Austin, attending the bar convention in Houston, and he needed to sign the bill before some other complication arose. Initial plans called for the bill to be flown to Sterling in Houston, but the possibility of early afternoon thunderstorms made flying impossible. The governor's executive secretary, Pat Dougherty, recommended they drive to Houston.[10]

Leaving the martial law zone, Colonel MacDonald drove south across the free bridge to the barricade manned by the Texas Rangers. The camps on the north and south banks of the river were within glowering distance of each other, and MacDonald wanted no misunderstandings between the armed men.[11]

Just in front of the barricade was a chalk line across the highway, scratched there by one of the rangers and supposedly marking the official Adams-Onís boundary line between Texas and Oklahoma. MacDonald was careful not to step over it when he introduced himself to Capt. Tom Hickman. The two men spoke for some time but, according to Hickman, battlefront conditions were not discussed.

In a later report, MacDonald said the rangers were dug in and didn't appear to be going anywhere soon.

"I'm sick of this dilly-dallying around," Governor Murray told a group of reporters on Friday afternoon. "The people on this side and in Texas might just as well get together now and have free bridges."[12]

The governor was adamant against toll bridges and toll-bridge investors, declaring they had made their money back many times over. He said the state of Oklahoma would ignore requests for improvement or reconstruction of the connecting roads that his highway department workers had destroyed.

By that afternoon, Governor Murray knew for sure that the Red River Bridge Company planned an action of its own. Since Monday, Joe W. Bailey had been shopping courtrooms, trying to get on the docket of a judge who would be willing to hear Bailey's request for a writ of mandamus requiring Oklahoma to open the roads to his client's bridge. Every judge in the Eastern Oklahoma District—most of them long-time Murray friends from the statehood convention days—recused himself because of conflict. Finally, Judge R. E. Lewis of the US Court of Appeals of the Tenth District in St. Louis sent for Judge Colin Neblett of New Mexico. Governor Murray was well aware that Neblett was flying to Oklahoma to preside in Muskogee on Saturday morning.[13]

"We're not bothering their toll bridge," he said. "They can travel back and forth on it until they are gray." Then, after a pause and a sly grin: "But they won't cross on this side of the river."[14]

With the United States Constitution as his sword and the National Guard as his buckler, Alfalfa Bill Murray showed he was ready to go to battle on behalf of Oklahomans, states' rights, and a free crossing over the Red River.

As is often the case in war, the frontline troops weren't privy to the grand strategy; they weren't seeing the big battlefield map. By Friday afternoon, they were seeing a very hot, very boring deployment. They had little to do but walk their post, talk to the curious children who gathered around them, pull grass burrs off their trousers, and swat at mosquitoes.

As cars entered the martial law zone, guardsmen waved them to a stop with small red flags. Gawkers, or tourists who missed the signs, were turned around; residents with passes (or post office mail drivers) were allowed to proceed. If a driver said he was on official business, the soldier would ask the visitor to get out of his car, then escort him to an officer. On the whole, the soldiers were polite small-town young men who did the job expected of them.[15]

Occasionally, a big shot would visit; two state officials with a direct interest in the bridge war arrived midafternoon. W. C. Lewis, the assistant Oklahoma attorney general, represented Oklahoma in much of the Red River Bridge Company litigation. Robert A. Keller, an attorney, was counsel to the Oklahoma Highway Department. He had prepared the legal brief on which Governor Murray based his decision to send the National Guard to the Red River. The men were inspecting the deployment before driving to Muskogee. They would be observing Joe W. Bailey's appearance in court the next morning.[16]

Local men, women, and children—small farm families, mainly—remained indoors during the early hours of the day, unused to having armed military men so close by. Gradually, though, as the day went on, the children left their houses to see the National Guardsmen up close. The boys—some in overalls, some in hand-me-down denims, and all of them barefoot—came close enough to reach out and touch the soldiers' war surplus cotswool shirts or the shiny brass boot eyelets. The first brave boy, the biggest boy, asked a question; then all the boys asked questions. Are your guns loaded? (No, admitted a young private.) Have you ever been on a tank? (Yes.) Can I wear your hat? (No.) The girls, generally, stayed behind the boys, shy and just out of reach.[17]

It was late afternoon before the grown-ups came down to the sandy road, supposedly to gather up the young ones but actually to look over the soldiers. The adults were lean and narrow people, mostly bone and muscle and not much more; their clothes were rough and homely and as suited for hard work as the folks who wore them. The farmers introduced themselves to the guardsmen, holding out plow-callused hands to shake. They shared tobacco, papers, and plugs while exchanging family names, expecting each to know or have heard of some branch of the other's family there in the county. Like the girls, the women stood back and silent, balling chapped hands into fists and pushing them deep into the pockets of their faded wrapover apron dresses, all the time hoping their men wouldn't invite the soldiers inside for a supper that would be spare enough as it was.

Before sundown, the farmers said their goodbyes and followed their wives home, promising to return after supper and the animals were fed.

Jake Loy, Pat Dougherty, Sidney Benbow, and T. B. Warden, a member of the Houston Chamber of Commerce, arrived in Houston at eight o'clock that night in Warden's car after a rain-soaked drive from Austin. A highway department auto followed them to provide transportation back to Austin.[18]

The group made the two-hundred-mile drive on improved state Hwy. 287 in less than five hours, a respectable pace given the weather. Loy was determined to get Governor Sterling's signature on the bill passed that morning by the legislature, but now—somewhere near downtown—they were lost.

The governor was due to speak to the convention of attorneys during their banquet at the San Jacinto Inn. Pat Dougherty knew the way to Sterling's home on Galveston Bay, but wasn't exactly sure how to get to the landmark restaurant. Warden had a booklet of strip maps, and was trying to locate the correct ones by the light of a streetlamp. Loy and Benbow, both on their first visit to Houston, had nothing to contribute.

Eventually, the party made its way out of Houston, through lush Pasadena and the refinery lights of Deer Park, to road signs marking the way to San Jacinto Inn, located alongside the Houston Ship Channel. At the restaurant, the banquet was in full swing, with huge platters of shrimp, crabs, and fluffy biscuits laid out family-style in front of the lawyers, but Sterling had not yet arrived. Another shuffling of the map strips set them on a course to the Sterling home in La Porte. They reached the residence in less than thirty minutes, but the housekeeper there told them the governor had already departed for the restaurant. They had passed him on the road in the dark.[19]

When the road-weary party returned to the banquet, they found Sterling at the head table, minutes away from giving his speech. The Fat Boy strode to the podium without an introduction and asked Sidney Benbow to bring the bill forward. In front of scores of cheering attorneys and two US senators, Gov. Ross S. Sterling signed the bill and returned it to Benbow to be filed with the secretary of state.

Sterling was still grinning from the podium when a diner called out, asking whether the governor's signature would allow the opening of the Red River bridge the following day.

"My dear sir," he answered, "Federal Judge T. M. Kennerly is better qualified to answer that question than I am."[20]

Dougherty would remain at the restaurant for the speech. He had been invited to spend the night at Sterling's home; they would work on other state business over the weekend. Warden left the banquet to retrieve his car; he would return to his home in Houston that evening. Benbow and Loy got into the highway commission automobile. The driver would drop Benbow at a Houston hotel to prepare for an appearance in federal court in front of Judge Kennerly the next morning. That left the highway department employee who would drive an exhausted Jake Loy, alone, back to Austin.

As Governor Sterling was signing the emergency legislation in front of an audience at the San Jacinto Inn that evening, the Dallas Chamber of Commerce was hosting drivers of the US Hwy. 75 Association in the banquet room of the Baker Hotel. The motorcade had arrived that afternoon and would depart for Sherman, Denison, Durant, and Tulsa early Saturday morning on its way to Winnipeg.[21]

Much of the conversation at the banquet involved the closed Red River bridges, the presence of National Guard troops at the Red River, and the effect this might have on the efforts of the association to promote their north-south interstate route.

Members of the association were businessmen from the cities through which US Hwy. 75 passed. Most of them had seen (or heard from parents) how towns bypassed by the railroads in an earlier time had withered and blown away, and they had no intention of that happening to their towns. The routes of the earlier named auto trails—Lincoln Highway, Meridian Highway, Bankhead Highway, Jefferson Highway—were established by individuals, promoters wanting to assure that their town wouldn't miss out on the money spent by tourists and freight haulers. But now—even as they dined that night—the US Bureau of Roads was ditching the old named routes and devising a new route system of numbered federal highways. Those numbered highway routes would receive more federal resources and be improved faster than would the intrastate roads. Members of the US Hwy. 75 Association were wary of anything that might threaten their numbered status.

Would the drivers have to admit their Galveston-to-Winnipeg route was still incomplete? If so, would the road bureau backtrack on US Hwy. 75 and change the route? How would the news of armed soldiers and pistol-shooting Texas Rangers at the Red River bridge threaten the future of each member's town?[22]

The diners would find no answers that night. They would have to wait to see what the next day would bring.

The Texas Rangers had been manning the bridge barricade for more than a week, and they were good at finding their entertainment where they could. After supper and sundown Friday night, Hickman fetched a wind-up phonograph and an assortment of records out of their equipment trailer. As usual, there were dozens of visitors to the bridge and ranger encampment, and Hickman took song requests.[23]

Soon, the Texas Rangers and their visitors were singing along to popular tunes and cowboy ballads, their voices echoing down the dark riverbed. "Moonlight Savings Time," "Sleepy Hollow," and (of course) "Red River Valley" were crowd favorites.

A group of girls from Denison begged Hickman to open the bridge for dancing, but the ranger vetoed the idea. As a consolation, he sang them a Mexican love song in a voice that was deep, loud, and mostly in tune.

The singing continued late into the night as the rangers waited to see what the next day would bring.

On the north bank of the river, residents of the military zone came out of their homes for more visiting with the guardsmen. The soldiers hung kerosene lanterns from trees so no one would trip on clumps of the plowed-up approach road. One farmer brought a watermelon, and he sliced it and shared it around.

The women gathered with their neighbors to talk; the children played games in the brush and chased lightning bugs.

In small groups, the men and the soldiers squatted on their hams in the jittery half-light while they talked about farming, livestock, crops, equipment, and their experiences at the time of the Great War. All of them—and especially the farmers near the river—had been touched by hard times. They speculated for a while about when the bridge would open so that they could get crops to Texas markets without complication, though some were so choked with worry they couldn't speak of it at all.[24]

There were no answers that night. They were all waiting to see what the next day would bring.

Shortly after midnight, in a highway department automobile returning to Austin from Houston, state senator Jake Loy lay across the back seat fast asleep. Loy had done as much as he could for his Denison and Sherman constituents.

Whatever the next day would bring would begin in courtrooms in Houston and Muskogee.

CHAPTER 12

If It Takes All Summer

THE NEXT DAY began at five thirty in the morning for Judge T. M. Kennerly of the US District Court in Houston when he received a phone call at home from an agitated Joe W. Bailey. Bailey, calling from Muskogee, made a last-ditch appeal to Kennerly to delay the injunction hearing scheduled in Kennerly's court later that morning.[1]

Bailey explained he had a hearing scheduled in Muskogee that morning. A favorable ruling in Muskogee might affect Judge Kennerly's ruling about the current injunction that was keeping the free bridge closed. In addition, Bailey argued, his client had never agreed to the waiver of the injunction in return for the right to sue Texas, despite what the Texans might be saying. Please allow my client the chance to be heard before you allow the free bridge to open, Bailey implored.

Judge Kennerly was noncommittal—though he may have reminded Bailey that his client had been granted the initial injunction as a result of Bailey's ex parte appearance before the court. In the end, Kennerly promised only that Bailey's client would have a chance to be heard.[2]

Lute Loy was another early riser. As chairman of the Sherman Chamber of Commerce committee dealing with the US Hwy. 75 Association caravan, it was Loy's job to lead participating drivers into Sherman, through Denison, and over the new Red River bridge to Durant later that morning. Loy had received telegrams from brother Jake and from Governor Sterling the evening before, telling him they were doing everything in their power to open the route. At seven o'clock on Saturday morning, July 25, however, the gleaming new free bridge was still barricaded by Texas Rangers and the toll bridge was blocked by the Oklahoma National Guard.[3]

Lute Loy was due to muster up a contingent of area businessmen at the courthouse shortly and lead them in their automobiles to Van Alstyne, a small town on US Hwy. 75 just south of Sherman. There, he and the local drivers—including a twenty-piece municipal cadet band—would meet up with the main motorcade arriving from Dallas.[4]

Whether he would lead them due north toward the Red River and the new bridge or on a bone-jarring roundabout detour through Preston and Madill, he didn't know.

Sidney Benbow was up early, too. The assistant Texas attorney general was having breakfast in his Houston hotel dining room, reviewing the motion he would present to Judge Kennerly later that morning.

At the original hearing, Judge Kennerly ruled that the contracts between the Red River Bridge Company and the state of Texas were valid and enforceable. Texas' refusal to pay the bridge company under the terms of the contracts left it with no legal recourse. The company was prevented by law from suing the state for the monies allegedly due it. The judge therefore restrained the state from opening the free bridge (and ruining the toll bridge company's business) until Texas settled up or until the toll bridge company found another way to collect.[5]

Now that the Texas legislature had passed a bill allowing the Red River Bridge Company to sue the state for its damages, Benbow would argue, there was no more need to deprive the citizens of Texas (and Oklahoma) of the benefit of their new toll-free bridge. That should be enough reason for Judge Kennerly to drop the injunction and let the traffic roll across the Red River.

Little Joe had also contacted Benbow, asking that his appearance in Kennerly's court be delayed. The conversation was a short one; the assistant attorney general said he intended to ask for an immediate ruling on the matter.[6]

Benbow put his papers—including a certified copy of the legislative action signed by Governor Sterling—back in his briefcase and prepared to leave the hotel. He was due in Judge Kennerly's courtroom at nine thirty that morning.

Joe W. Bailey was due in the Muskogee courtroom of Judge Colin Neblett at the same hour.

On the north bank of the Red River near the entrance to the toll bridge, the Oklahoma National Guardsmen were being served a breakfast of eggs, potatoes, ham, and coffee out of the back of a truck. Their meals were prepared in Durant, were brought by truck to the martial law zone three times a day, and were predictably less than appetizing. The meals weren't all bad, however. The women of Durant had taken to their kitchens, supplementing rations with hot dishes, soups, breads, pies, cakes, and other homemade treats.[7]

On the south bank of the river, still guarding their barricade, the Texas Rangers ate meals prepared by a Denison diner. Waitresses from the diner vied with one another for the privilege of delivering the meals to the rangers at the bridge.

There was no reason for the soldiers or the rangers to think that Saturday would be anything other than another boring day, so boring, in fact, the rangers sent a boy across the bridge to deliver a message to the Oklahoma soldiers: "Now you know how much fun we're having."[8]

As usual, *Denison Herald* reporter Claud Easterly arrived early at the bridges to catch up on any events that may have occurred overnight. The Texas Rangers and Colonel MacDonald of the National Guard told of a quiet evening by the river, and both sides were expecting a slow day. The action, they said, would be happening in Houston and Muskogee, not at the bridges.

This was particularly good news for the young reporter. For two weeks he and his newspaper had been covering one of the biggest breaking stories in the nation, and it was thrilling. But Easterly absolutely had to have the afternoon and next few days off work. His wedding ceremony was scheduled for five o'clock that evening.

The bride was a high school sweetheart. Easterly had gone directly into the newspaper business after high school; Ruthie Davis had attended Southeastern State Teachers College in Durant and had graduated from Austin College in Sherman the previous January. J. Lee Greer, general manager of the newspaper, and the rest of the staff had staged a surprise party for the couple at Woodlake earlier in the week, but the wedding would be a small affair, held in the bride's parents' living room with only immediate family in attendance. Time off for the honeymoon had been cleared with his boss, and Easterly had been warned—by everyone—not to be late to the ceremony.[9]

After his early visit to the river, he returned to the newspaper office to write a column of story bits that would run under a "Sidelights at Battlefront" standing headline. Thankful for a slow news day, Easterly planned to wrap up at the office and be home at noon for lunch with his parents.

Joe W. Bailey was angry when he walked into the courtroom in Muskogee at nine thirty that morning. He was angry at having to step into yet another courtroom to protect his client's property; he was angry that the governor of Texas had reneged on a contract intended to end their legal contests, resulting in even more litigation; and he was angry at the governor of Oklahoma for having destroyed his client's property and business, shredding the approach road and preventing the company from collecting tolls.

Until that morning, Bailey had been the primary legal adviser to the bridge company for more than two years. His father had represented the company for two years prior to that, and he had watched his father draw his last breath—literally—in a Sherman courtroom in defense of the Red River Bridge Company. The state had no business treating a legal enterprise in such an arbitrary, overbearing, destructive manner.

Acid dripped from the words of his petition and from the words he would speak to the judge.[10]

"Your honor, I am seeking injunctive relief against this final theatrical move of martial law," Bailey told the court. His petition named twelve individuals, including Adjutant General Charles F. Barrett, Burt McCaleb and his divisional field engineers, the Oklahoma Highway Commission members, and twelve seemingly random individuals from the Durant and Colbert areas.

Bailey entered into a lengthy history of the dispute between the toll bridge companies and the states of Oklahoma and Texas, and Judge Neblett listened patiently. The attorney described the contracts that should have ended the litigation and accused Governor Sterling and the Texas Highway Commission of "fraudulent intent" and "fraudulent concealment." Though not a party to the contracts, Oklahoma benefited from them, he said, and the state highway commission, the engineers, and everyone else named in his petition "breached or assisted in breaching each and every term and provision" of those contracts to his client's detriment.[11]

The federal court in Texas granted an injunction to close the new bridge until our contracts were settled, Bailey told the court. "Governor Murray evolved a scheme to nullify the action [of the court] by so-called executive order to open the parallel free public bridge." The defendants "joined in with the scheme and some of them, while bearing arms, tore down the barriers ordered" by the court. Some of the defendants, "acting under the direction of Burt McCaleb, proceeded to cross the bridge into the jurisdiction of the Eastern Division of Texas and publicly announce that the free bridge was open for travel by all who cared to pass over it." Then, using a tractor and also under the direction of Burt McCaleb, "some of the defendants destroyed at least 1,200 feet of roadbed" leading to the toll bridge," doing so to deprive his client the right of any compensation for its property.

Bailey then named the twelve individuals from Durant and Colbert. They "direct traffic ordinarily intending to travel over the toll bridge to proceed by other different, more circuitous and less-traveled, and worse-conditioned roads" for the purpose of starving out the toll bridge company. The attorney asked that they, as well as the other defendants, be prohibited from giving directions around the toll bridge.

Bailey entered into evidence a copy of Governor Murray's "purported proclamation of martial law," telling the judge that "the governor's intention, as publicly stated, was to forestall any contemplated orders of this court to remove obstructions leading to the toll bridge." General Barrett "was ordered to maintain military control against all interference." Murray's orders were still in effect, Bailey reminded the judge, "for the expressed purpose of preventing this court from taking any action on the premises."

Judge Neblett appeared somewhat amused by the events Bailey described, leaning back in his chair and smiling from time to time. But when Bailey described Governor Murray's preemptive deployment of the National Guard, Neblett came to attention.

Bailey requested three things from the judge: First, he asked that all parties named in the petition be enjoined from blocking or discouraging access to the toll bridge in any manner. He asked that the judge restrain the state of Oklahoma from allowing traffic to cross the free bridge by erecting a barricade on the Oklahoma end of the free bridge (and allowing the toll bridge company to maintain the barrier). Third, he asked for damages of a quarter million dollars to repair the bridge approach road and for revenues lost during the closure.

Before Judge Neblett adjourned, W. C. Lewis, who was observing the proceeding and making notes, asked to speak. Smiling and confident, Lewis briefly reminded the judge that the state of Oklahoma and its officers were immune from lawsuits such as this.

Bailey had been expecting that argument, and he jumped to his feet, citing precedent for injunctions and damage actions against state officials. "Persons usurping power, hidden behind the skirts of a sovereign state" can be made liable to lawsuit, he said.

Judge Neblett adjourned, telling spectators he would deliver his decision at a later time. He suggested that the parties remain close to the courthouse.

Four hundred seventy-five miles south, also at nine thirty Saturday morning, Sidney Benbow's appearance in Judge Kennerly's Houston courtroom was less intense, but no less important to events along the Red River.

Benbow was asking Judge Kennerly to dissolve the injunction he had put in place three weeks earlier. There was no reason to repeat the facts of the case; Judge Kennerly knew them. The judge asked only for a copy of Benbow's petition and a copy of the related bill signed into law by Governor Sterling the previous evening.

Judge Kennerly adjourned, telling Benbow he would deliver his decision at a later time. He suggested that Benbow remain close to the courtroom.

The first visitors to the free bridge arrived early; by ten o'clock hundreds of local men, women, and children milled about near the rangers' barricade and the grassy bank, drawn by curiosity about what the day might bring. The rangers spoke with the sightseers, but kept the visitors away from the barricade and off the bridge. Eventually several sheriff's deputies turned up to direct traffic and help control the crowd.[12]

The babel of the crowd and usual outdoor sounds at first drowned out the racket, but an unfamiliar buzzing, popping noise grew in volume until people were turning to their neighbors, asking what it might be. One sharp-eyed man, looking upward, pointed downriver and people turned to follow his arm.

It was an airplane, a war surplus Jenny with open cockpit, flying slow and low enough for passenger and pilot to exchange waves with the crowd. Several airplanes would fly over the river that weekend, some no more than twenty feet above the riverbed, hopscotching over the steel bridge trusses like an aerial roller coaster. Enterprising flying companies and itinerant pilots were offering sightseeing tours by air over the battlefront because, as one reporter noted, airplanes were the only mode of transportation not prohibited from crossing the river.[13]

Judge Kennerly didn't keep Sidney Benbow waiting long. The judge summoned the assistant attorney general back to the courtroom for his decision.

"It is ordered that the motion to dissolve the temporary injunction, granted July 9, 1931, in this cause be set for hearing at Houston on the third day of August, 1931," Judge Kennerly said. "Pending such hearing, the enforcement of said injunction is hereby suspended."[14]

In fewer than thirty seconds Judge Kennerly informed Sidney Benbow that he could open his bridge.

Benbow's first phone call was to Governor Sterling, who had been standing by at his Houston-area home for news. Sterling then asked his secretary to send a telegram to the Texas Rangers at the bridge, telling them to remove the barricades and open the free bridge. As the governor placed calls to Austin, Benbow sent his own wire to Denison.[15]

In Denison, reporter Claud Easterly learned of Kennerly's order (probably by way of a tip from the Western Union operator). After telling his editor what he knew, Easterly dashed off to the bridge with notebook and pencil, knowing he would miss lunch with his parents, but sure he could get back to town in time to dress for the five o'clock wedding. Easterly was a newspaperman; his parents would understand.

Western Union handed Governor Sterling's telegram to Denison mayor Clarence Scott, who raced off to deliver it to the rangers. Accepting the telegram, Captain Hickman opened the envelope and saw it was signed by the governor. Afraid of a hoax, Hickman tried to reach Governor Sterling in Houston. While waiting for that call to go through, Hickman heard from Gibb Gilchrist, chief highway department engineer, telling Hickman he could instruct highway workers to remove the barricade. Still Captain Hickman delayed taking action. Finally, Hickman reached Adjutant General W. W. Sterling, who told him he had spoken with Governor Sterling personally. General Sterling ordered Hickman to open the bridge.[16]

"This is not the first boundary dispute I've been in," Hickman told the general. "I want to know that it's coming straight."[17]

Only after receiving direct orders from his superior officer did Texas Ranger captain Tom Hickman give the order to dismantle the barricade and let the traffic flow.

News of the opening spread through the crowd as the highway department workmen arrived at the barricade with their tools. Hundreds of spectators pressed forward, cheering, waving hats and handkerchiefs. Hickman and the other Texas Rangers had to hold back the swarm as some spectators tried to help the workers pry and wrench heavy timbers apart.[18]

"Stand back," Hickman ordered the crowd. "Let them work and earn their money." Several men were able to dodge the rangers, pull boards from the barricade, and toss them down the embankment.[19]

Some in the crowd ran to their automobiles, lining up alongside the road in hopes of being among the first to cross the bridge. They sat behind steering wheels, honking horns, until the din was so great it was hard to hear.

The car nearest the fast-disappearing barricade was a new Buick sedan driven by R. G. Gresham of the Denison Chamber of Commerce. With him was B. J. Lindsey, the Red River Bridge Company shareholder who had championed the new free bridge. As Hickman prepared to wave them forward through a gap in the blockade, something shot past him at high speed going north. The Texas Ranger whirled around in time to see a boy pedaling his bicycle hell-bent-for-leather toward the Oklahoma end of the bridge.

The first vehicle to cross the newly opened toll-free bridge over the Red River was a bicycle ridden by fifteen-year-old Roy Epps of Sherman. People at both ends of the bridge clapped and cheered him on.[20]

Hickman signaled the Buick to proceed across the bridge, then jumped onto the running board to accompany Gresham and Lindsey and inspect the roadbed for any debris or obstruction. Before starting across, Hickman looked south to see a group of decorated automobiles topping the rise from Denison and driving toward the bridge—headlamps flashing, horns blaring, passengers waving, and the sound of a band playing "Dixie." It was the Highway 75 Association caravan, led by a grinning (and relieved) Lute Loy. The ranger captain signaled Loy and the motorcade to follow him across the bridge.[21]

The thirty-car caravan took its time crossing the new bridge, as drivers and passengers admired the width of its traffic lanes, its sturdy roadbed, and the silvery overhead trusses that reflected flashes of midday sunlight. This section of roadway represented just what the Highway 75 Association wanted to see: an unbroken ribbon of paved highway and thoroughly modern bridges stretching from the snows of Saskatchewan to the oleanders of Galveston. They believed US Hwy. 75 would be America's premier north-south route.

Halfway across the bridge the motorcade passed an automobile driving south, the first automobile to drive the free bridge from Oklahoma to Texas. Stuffed inside the older Ford were Mrs. Lucille Diffey and her three daughters, Mrs. Maude Wilson and her three children, and Miss Marie Bacon, all from Colbert. Hearing that the bridge was to be opened, they had chosen to drive to Denison for a day of shopping.

At the Oklahoma end of the bridge, Hickman stepped off the running board and posed for pictures before hitching a ride back to the Texas side. There, he and the Texas Rangers of Company B packed up their gear and drove away from the Red River bridge to their next assignment.[22]

Beginning at noon that day a steady stream of automobiles were being driven north and south across the free bridge, bypassing the closed toll bridge, avoiding the seventy-five cent charge for traveling the thousand yards across the Red River.

As the Highway 75 Association caravan crossed the new bridge amid cheers and waves on its way to Durant, federal judge Colin Neblett in Muskogee was taking the bench to deliver his decision on the motion Joe W. Bailey argued that morning.

Judge Neblett had listened to Bailey's earlier recitation of events at the bridge with a bemused smile, but he was solemn as an undertaker when he read his order, perhaps recognizing the conflict that his words might trigger.

Bailey had successfully argued the damage to his client's property, the judge said, and he had shown how the parties named in the motion continued to diminish the value of the toll bridge. He granted Bailey a hearing on the matter of damages on August 7.

In the meantime, "plowing up, tearing up, destroying or erection of barriers" on the roads leading to the toll bridge were forbidden, the judge said.[23]

General Barrett, the National Guard, the highway commission, and Oklahoma highway department employees were required to permit the Red River

Bridge Company to replace the toll bridge approach road "back into condition for traffic and to tear down and remove any barriers."

The order also prohibited the guardsmen and, by name, the twelve southern Oklahoma business operators from "dissemination of any information at any point or place" that might draw traffic away from the toll bridge.

Judge Neblett paused and looked directly at Assistant Attorney General Lewis in the gallery as he read the last section of the order. "If the National Guard or other state agency interferes with the peaceful performance of these acts," Neblett read, "the federal marshal is to return immediately to this court and obtain warrants against all interfering persons." Still looking at Lewis, Judge Neblett said that the court might also choose to issue a contempt citation against Governor Murray, even though he was not specifically mentioned in Bailey's motion.[24]

The judge announced his intention to jail any party who resisted his order.[25]

Alfalfa Bill Murray was determined to resist the judge's order.

W. C. Lewis phoned Governor Murray with a report as soon as court was adjourned. Always the attorney, Murray wanted to read the order himself, and he instructed Lewis in how it should be delivered to him.[26]

Shortly after the conversation with Lewis, Murray walked to his residence and began throwing a change of clothes into a valise. With the free bridge open and the rangers withdrawn, the governor had accomplished what he had set out to do. But he was angry that a federal judge would presume to interfere in the military matters of the sovereign state of Oklahoma. According to Lewis, the federal court was ordering the National Guard to stand down, and Joe Bailey had tied General Barrett's hands by including him by name in the suit. The governor was on his way to the Red River with the intention of taking command of his troops, he said, and "to live the life of a soldier."

Hearing that the governor was going to war, his worried valet rushed from the room and as quickly returned with an ancient long-barreled revolver wrapped in a handkerchief.

"I never carry a gun," Murray told the valet, "but I'll take this one." The valet placed the big pistol and a box of ammunition in Murray's suitcase.[27]

Less than an hour after Judge Neblett read his order in Muskogee, Alfalfa Bill was in the back seat of a car driven by A. R. Losh, chief engineer of the highway department, and on the road speeding south toward the river. Murray took off his shoes, draped his coat over himself, and closed his eyes for a nap.[28]

The nap didn't last long. On the road through Norman, a policeman pulled over the governor's car. The Norman police force was on the lookout for Murray, the patrolman said, because they had received an urgent message from W. C. Lewis. Lewis would meet the governor in Ardmore, and the policeman passed along details.

Continuing the drive south, Losh was forced to stop at the booth of a toll bridge over the Canadian River between Lexington and Purcell. This was the only remaining privately owned toll bridge in Oklahoma, and the state had filed a lawsuit to acquire it by eminent domain.

"Sure, they paid the toll," the bridge keeper told a reporter after Murray's car had passed. The bridge keeper had recognized Alfalfa Bill in the back seat. "The governor didn't pay it himself, but his driver did."[29]

At Ardmore, Losh drove to the municipal airfield and pulled the car up to a hangar there. In minutes, a plane landed and taxied to the governor's car. W. C. Lewis climbed down from the passenger cockpit, wearing a leather flying helmet and carrying his briefcase. The assistant attorney general had been forced to fly to Ardmore from Muskogee in order to catch up with the governor on his way south. Lewis got into the car with Murray, and the two of them reviewed Judge Neblett's order.[30]

Lewis had accepted service of the restraining order on behalf of the state and all named parties. He warned the governor that Judge Neblett appeared deadly serious about compliance, and probably wouldn't waste any time sending US Marshals after anyone who disregarded it.

Murray was defiant. "He can't cite me for contempt," Murray said. "Let him try."[31]

As Lewis and the governor conferred in the back seat, Losh continued the drive toward Durant and the bridges, stopping briefly in Tishomingo for Murray to place some phone calls. Governor Murray knew his law, knew his Constitution, knew the ramifications of Neblett's order, and knew what he was going to do about it.[32]

Joe W. Bailey was still worked up over the events of the past several days, but he was about to get some satisfaction for the toll bridge company: he would take back his client's bridge from the hands of the Oklahoma National Guard.

Bailey departed from Muskogee with certified copies of Judge Neblett's order and drove south toward the Red River, picking up speed as he crossed onto hard-surfaced US Hwy. 75 at McAlester. He crossed the Red River into Texas on the free bridge—though it probably irked him to do so—into Denison at midafternoon, where he met two armed deputy US Marshals who had driven up from Dallas. Bailey and the federal marshals—followed by a gang of a dozen or more newsmen—walked across the toll bridge to the Oklahoma side and asked the first National Guardsmen they encountered to summon the commanding officer.[33]

Shortly after, Lt. Col. John A. MacDonald arrived at the bridge wearing full dress uniform: khaki blouse, service coat buttoned to the neck, breeches, and puttees, all topped by a perfectly crimped campaign hat, placed so that the hard felt brim was absolutely horizontal. He introduced himself to the three men, apologizing that he was unable to invite them to his headquarters.

"This bridge is closed, as you know," MacDonald told them with a smile.

Joe Bailey was not smiling when he informed MacDonald he was there with an order signed by a federal judge. The order required that MacDonald and his men open the toll bridge and leave the area immediately. He asked MacDonald, a fellow attorney in civilian life, whether he would accept service of the order.

"You can leave a copy with me, if you wish," MacDonald told Bailey, "but I will ignore it."

Bailey was incredulous. "Do you mean to say that you intend to defy a federal court order?" Bailey asked.

"Yes," MacDonald replied. "I am taking orders from only one man, and he is the governor of Oklahoma."

Red-faced and quivering with rage, Joseph W. Bailey Jr. turned on his heel and returned to the Texas side of the bridge without leaving the order.

Judge Neblett also departed Muskogee shortly after adjournment, eager to return home to Santa Fe as soon as possible. He was scheduled to fly from Tulsa to Albuquerque, with a stopover in Amarillo, where reporters were waiting to speak with him.

Neblett seemed amused over the entire bridge war and chuckled when a reporter told him that Governor Murray was en route to the Red River to take personal charge of his militia.

"I don't think he'll openly defy the federal order," Neblett said, "but if he does I'll cite him immediately for contempt of court."[34]

Neblett's words—stripped of the wry humor with which they were spoken —went out on the wires and would quickly reach Governor Murray.

Claud Easterly missed the toll bridge confrontation between Bailey and MacDonald. In one of the wisest choices of his newspaper career, the young reporter turned his back on a breaking news story of national interest and returned home to dress for his wedding that night.[35]

Easterly was enough of a newspaperman, however, to stop by the newspaper office and leave his notes with the editor.

At seven that evening, as Easterly and his wife were exchanging vows, Alfalfa Bill Murray arrived at the first barricade to the toll bridge in the martial law zone. A National Guard private flagged down the governor's car and was struck speechless when Governor Murray rolled down the back seat window and asked to speak with Gen. Charles Barrett.

Murray's first task was to relieve Barrett of duty, not because of any error or blunder, but to take the military man out of the line of fire of any further legal action. Murray asked Barrett to assemble the soldiers so he could tell them of the change in command.

"I am taking complete personal charge," the governor told the guardsmen and a group of reporters who had heard of Alfalfa Bill's presence. "I will take all responsibility for everything that is done."[36]

Barrett wasn't the only one to be taken off the legal hook. Murray sent Losh and Lewis to Durant to formally serve Burt McCaleb and his highway crew with notice of the injunction and relieve them of their highway department jobs. Lewis then commissioned each of the (temporarily) unemployed men into the National Guard and told them to report to Governor Murray at his new headquarters. The baffled highway men departed for the river, still shaking their heads and wondering what had just happened.

Murray told the gathered reporters that every man in the martial law zone along the toll bridge approach roads was a military man and under his direct command.[37]

That business out of the way, the governor—still dressed in his white linen suit and straw hat—walked down the approach road toward the toll bridge, stopping to exchange a few words and shake the hand of every soldier along the way.[38]

It was supper time, and an evening meal of liver and onions was being served from the back of a mess truck. Murray joined the line and took a large slice of liver and several stewed onions. To the amazement of the guardsmen (and the delight of accompanying newsmen), Alfalfa Bill sat down with the soldiers for the meal.

We're going to "hold the fort," Murray told them, against efforts to enforce the injunction opening the toll bridge or any new orders closing the free bridge again. He promised to stay with the soldiers in the field "night and day" until all threat was gone and martial law was lifted.[39]

The governor rejected an offer to take a hotel room in Durant for the night. "I'm looking forward to camping out," Alfalfa Bill said as he finished his meal. "I'm used to that kind of life."[40]

To prove his point, he asked a private to set up a cot in the abandoned brick headquarters building near the entrance to the toll bridge. Followed by reporters, Murray made one more tour of the military camp before turning toward his quarters, fully intending to turn in for the night.

Well before ten o'clock Saturday night, Claud Easterly and his new bride—she dressed in her traveling outfit of a marine blue suit with white blouse, white wool turban, and white footwear, he in a dark suit—departed Denison for their honeymoon. They planned to spend the night in Dallas and leave the next morning for a three-day stay at the posh Baker Hotel in Mineral Wells.[41]

Through the window of the brick maintenance building in which Alfalfa Bill Murray would spend the night, he could see the electric lights that MacDonald's men had installed along the length of the toll bridge. "From the distance, the toll bridge resembled a carnival, so brilliantly was it lighted," according to one observer. Sounds, too, carried up to the building from the free bridge a half mile upstream. A soft summer breeze brought the noise of motors, horns, and jubilant shouts from the crowded free bridge.[42]

Governor Murray was likely not noticing the lights or the sounds; he was stretched out on his cot, shoes and coat off, answering reporters' questions. Always willing to take one more question, make one more statement, or give a lecture, Murray lit another cigar as he succumbed to the reporters' entreaties for more quotes about his bridge war strategy.[43]

He said he could see no prospect for an early settlement, and he repeated his promise to remain personally at the riverside military headquarters until the free bridge was opened permanently, "if it takes all summer." He didn't expect any real developments until the Houston and Muskogee hearings, scheduled early the following month.

"I had no idea they would go that far," Murray said about the federal court decision in Muskogee that morning.

"The character of this [decision by Judge Neblett] is so important under the Constitution of the United States that the bridge question sinks into insignificance," Murray declared, punching the air with his stogie. "The question is whether or not a sovereign state still possesses its powers under the Constitution."

Murray couldn't imagine that a federal judge would prohibit private citizens from giving travel directions, but he was particularly indignant that the judge would require the states to allow the toll bridge company to make use of state roads.

"The whole question is whether the power and rights of the corporations surpass those of the people," he said. "If I abandon this fight now I would be abandoning my duty to uphold the powers of the state."

At this moment Alfalfa Bill had every justification for declaring victory in the Red River Bridge War, gathering up his troops and returning home. The Texas Rangers were gone, the free bridge was open and traffic was flowing across it, and the toll bridge would have a hard time collecting any tolls with a (better) free bridge just a half-mile away. Ross Sterling was drafting his concession statement, and newspaper editors were composing editorials unanimous in crediting the Oklahoma governor for having opened the free bridge with his forceful actions.

Murray could have declared himself the victor of the month-long war and gone home.

But he didn't.

Instead, he recast the Red River Bridge War as a constitutional confrontation, a battle between a sovereign state and the federal government.[44]

CHAPTER 13

Cross Another Creek

DESPITE HIS late night with reporters, Alfalfa Bill rose early on Sunday morning. More newsmen had arrived overnight and they surrounded Murray as he walked out of the small maintenance shed shortly after sunup. (In news stories filed overnight, some reporters datelined their coverage as coming from the "Alfalfa Bill Barracks on the Red River Front.")[1]

Before walking to the mess wagon for breakfast, Murray told gathered reporters the story of a late-night visitor to his barracks. According to the governor, the visitor was a watchman from the free bridge who disclosed a secret plan to block the southern entrance to the free span once again. The purported plan, according to the watchman, involved Deputy US Marshals, acting under a new court order not yet revealed. The federal marshals would block the south end of the free bridge, closing it to traffic once again, and then enforce the opening of the toll bridge.[2]

Newsmen accepted this story of a midnight stealth visitor without challenge (although several of the writers had been sleeping on the ground outside the small maintenance building that served as Murray's quarters and had seen no one pass.) Apparently, the governor believed the bridge war was still on.

Ross Sterling, however, had conceded the battlefield. The Texas governor released a statement on Saturday night, and a reporter read it to Murray over breakfast Sunday morning.

"So far as I am concerned it is history now," Sterling had written. "I am very proud to note that the bridge was opened according to law and not by high-handed methods of state authorities."

Before Murray could comment, the reporter continued reading: "Personally I have the very kindest feeling toward the chief executive of Oklahoma and hope that the matter has ended to the satisfaction of all. The people are now using the bridge and everyone should be happy."[3]

Sterling had every reason to be happy. With the free bridge finally open, the Texas Rangers decamped, and Denison officials already beginning to revive their bridge opening celebration plans, Sterling seemed entitled to a huge sigh of relief that he was at last free of the tar baby that had stuck to him for the last month. Now it was Murray's turn to be bogged down by the river, by the bridges, and by the courts.

If Murray was bogged down, he wasn't showing it. Instead, he seemed more a combination guest of honor and party host; he was the center of attention and leader of the outdoor activities, too.

After a breakfast of buttermilk, biscuits, and bacon, Alfalfa Bill announced he was ready to inspect the toll bridge. He asked Colonel McDonald to form up the guardsmen into columns to "escort" him to the battle lines.

Murray usually disdained the pomp and formalities of his office, but he also had a publicist's eye for what would play in the press. The presence of dozens of newsmen, photographers, and at least one newsreel cameraman brought out the showman in the governor. An elderly visitor told the governor that morning that all the military activities at the encampment reminded him of Theodore Roosevelt in Cuba. "You know, every time Teddy came to a ditch or a creek he had General Wood stop to take a picture and send it back to the US papers," the visitor said.

Murray thought the story was hilarious and, for the rest of the day, when Murray halted the crowd of press people around him to pose for another picture, he would announce, "We've gotta cross another creek, boys!" The line brought laughter from the governor and reporters every time it was used.[4]

Murray's inspection trip to the toll bridge and back—a three-hundred-yard march—took more than three hours. "Posing Takes All Murray's Battle Time," one newspaper headline writer explained.

Alfalfa Bill wasn't about to let anyone forget his constitutional battle (and other politicians were eager to share the fight and the riverside spotlight with the governor). US Rep. Tom D. McKeown of Ada called on the governor at his barracks later that morning to declare some of the provisions of Judge Neblett's decision the previous day "preposterous."[5]

Enjoining local residents from using words or signs to divert traffic was in obvious conflict with the constitutional guarantee of free speech, the congressman fumed to reporters gathered around himself and the governor. McKeown promised a congressional investigation of federal judges and passage of a bill limiting the issuance of federal injunctions.

Several "peace delegations" called on Murray that day. Six men from Tishomingo—all wearing a version of Native costume that would have been more familiar to Hoot Gibson or Tom Mix than to anyone from the old Indian Territory—drove down to the river to wish their fellow tribesman good luck in his war with the white men. Murray took it all with good humor, and the men departed to join the steady stream of traffic crossing the new toll-free bridge upstream.[6]

Delegations from Denison, Durant, Sherman, and Gainesville arrived to thank Murray for opening the bridge. Each of the delegations included businesspeople who told Murray how much their businesses had suffered while the bridges were closed. The Texas Highway Department had posted its detour signs south of Sherman on US Hwy. 75, so tourists never had a chance to enter (and spend money in) Sherman and Denison. The mayors of Denison and Gainesville—each with his own further political ambitions—encouraged Murray to run for president in the upcoming election.[7]

(Murray declared he hadn't thought about it, but if he ever ran for another office he would "run for president or nothing.")[8]

Alfalfa Bill, playing to the cameras and press, shook hands with each visitor, thanking them for their support of his stand against a runaway federal government that most often favors the corporations instead of the farmers and others who need help the most.

It wasn't just notables who came to meet Governor Murray. An old Red River fisherman walked up from the riverbank into Murray's headquarters.

"A bunch of us down on the river have made up a pot to build a big sign by the free bridge," the fisherman told the governor. They wanted the sign to honor Alfalfa Bill for opening the free bridge. "Do you reckon the highway fellows would take it down?"[9]

A chuckling Murray said he thought not, but he glanced at Losh, the highway engineer, who said plans were already under way to erect a bronze tablet in honor of the chief executive at the free span.

Calvin Hannan was a Colbert man who grew cotton, corn, peanuts, melons, or whatever people were buying that season. He owned farmland two miles north of town and leased some acreage a few miles east. Hannan and his two sons hiked down to the toll bridge, "just to see what was going on."[10]

People said Calvin Hannan could make a friend of anyone he met, and he greeted Alfalfa Bill Murray like a Baptist deacon (which he was) welcoming a church visitor. Murray stood up from his camp stool as Hannan introduced himself and his boys. For twenty minutes the men spoke of farming, crop prices, roads, and bridges. Opening the toll-free bridge opened markets for Hannan; the farmer could compete on an equal basis with growers on the other side of the river.

Sunday was a day marked by laughter, wisecracks, cheers, and heartfelt appreciation as Murray—sitting in front of his barracks—met fans and admirers at the guard encampment.

Herald reporter Claud Easterly (who had cut his honeymoon short to return to Murray's riverside barracks) described an epidemic of "handshake fever" among Denisonians. The malady, he said, struck prominent citizens who boasted about shaking the hand of Alfalfa Bill Murray, "champion of free bridges and bitter enemy of the toll structure."[11]

"The bridge war seems to have raised the batting average of the governor's popularity hereabouts," Easterly wrote.

Reporters surrounded Murray, jotting down every word as he greeted old friends and reminisced about his days in Texas, Tishomingo, and Bolivia. He talked about trying to practice "cowboy law" in Fort Worth and the political wrestling matches necessary when chairing the state constitutional convention. His solders, particularly, were interested in the Bolivia stories, and he spoke at length about the hardships of his colony and the great engineering feats he had seen in South America.

For all that, it was his six months as governor that proved most eventful. "Already I feel like I've been governor of Oklahoma for a longer time than the five years I spent in Bolivia," he mused.[12]

The conversation broke up when afternoon mess was announced. The guardsmen were served chicken and dumplings and iced chocolate milk; Alfalfa Bill had ordered his favorite meal, fat bacon and cabbage, which the mess officer obligingly served.[13]

After dinner, a reporter asked Murray if he needed a nap, especially after his late night the night before. Giving it some thought, he said he wasn't much for daytime sleeping. He drew laughs when he said, "Right now I'm too busy with my army and with opening up the bridge to get much rest."[14]

Perhaps the day had tired him. Or, perhaps, Murray realized that his army of thirty-five National Guardsmen didn't need his help, and that the free bridge was already carrying traffic across the river. For whatever reason, Murray stood up, walked into his barracks, and began packing his valise.

Governor Murray's decision to leave the encampment that night was unexpected. And he had an even bigger surprise in store.

Before departing for Oklahoma City, Alfalfa Bill—with his corps of reporters in attendance—gave new orders to Colonel MacDonald. The governor ordered MacDonald to maintain the military zone at the toll bridge, but to allow any driver willing to pay the toll to cross the bridge without interference.[15]

Murray told reporters he feared that if he released his military power at the toll bridge until the matter was definitely settled, the bridge owners would take possession using federal marshals. That would place the governor in the position of attacking the federal government to close the toll bridge again if needed.

"I can tell you that they would already have taken it over if I hadn't sent the army to keep it closed," he said.[16]

Though the governor failed to mention it, by allowing the toll bridge to operate unimpeded Sunday night, Alfalfa Bill had assured compliance with all requirements of the injunction issued in Muskogee the day before. General Barrett had been relieved from command of the guard, highway department employees were instructed to not interfere with toll bridge access, and the National Guard was assuring the toll bridge would be open to any who cared to use it. Nevertheless, the governor warned of the abolition of state governments and "a setting up of centralized despotism if the federal district court attitudes in the bridge war were allowed to stand."[17]

"The free bridge is open now," Alfalfa Bill told reporters. "If people are fools enough to want to pay seventy-five cents to cross here, let 'em do it."[18]

With that, Governor Murray threw his valise into the trunk of his state-owned automobile, slid into the back seat, and was driven away from the Red River, back to his office at the capitol in Oklahoma City. By ten o'clock

Sunday night, not a single automobile had chosen to pay the toll to cross the reopened toll bridge.[19]

Traffic continued to pour across the free bridge on Monday morning, July 27; no motorists elected to cross the river using the toll bridge only to face the bulldozed northern approach road. If the free bridge were allowed to remain open, the Red River Bridge Company's bridge would soon be worth no more than its weight in slag.

Attorney Joe Bailey, still smarting from having been turned away from his client's bridge by Oklahoma National Guardsmen two days before, had few options once Governor Murray belatedly complied with the Muskogee injunction. Because of special legislation, the bridge company might eventually have its day in a Texas court; but that might be too little, too late. Judge Kennerly in Houston had voided the order that kept the free bridge closed, but scheduled a full hearing on the matter the following week. Perhaps the judge could be persuaded to act before then.

That morning Attorney General Allred picked up rumors that Little Joe Bailey would ask the Houston court to reinstate the injunction preventing the use of the free bridge. Closing the free bridge, Allred knew, would spark another round of barricades, Texas Rangers, and traffic snarls.[20]

Alfalfa Bill heard the same rumors. By noon Monday he had composed and transmitted new orders for the National Guardsmen stationed at the Red River, orders that threatened to re-ignite the Red River Bridge War. By two o'clock Lieutenant Colonel John MacDonald was redeploying his troops under the new orders, this time to protect the free bridge.

"An effort has been made to close the bridge against any traffic whatsoever," Murray's new executive order said. "There are rumors of other attempts to close the free bridge and approaches."[21]

Governor Murray's new order extended the existing martial law zone to include the free bridge and the north and south approach roads leading to it. MacDonald set up two additional sentry posts: one on the Oklahoma side of the bridge, the other about eighty feet south of the bridge at the first cut-bank, a place where the Texas Rangers drew their chalk line marking the Adams-Onís Treaty border between the two states.

The governor's order to MacDonald was "to keep both bridges open to traffic, whatever the cost."[22]

The new deployment was largely meaningless. Little Joe Bailey made no petition to the federal court in Houston; newsmen failed to unearth any other credible threats to close the bridge. Governor Sterling—stung by public reaction to his earlier efforts—largely ignored Oklahoma's latest armed incursion, and Texas adjutant general W.W. Sterling laughed off rumors that Texas Rangers would again be dispatched to the bridges.[23]

Residents of Durant, Denison, and Sherman also paid little notice to the sentry posts on the free bridge—they were busy organizing their long-delayed bridge-opening celebrations. The mayor of Denison called a Tuesday night meeting of the celebration committee to dust off plans for the two-county picnic; a pavement dance; terrapin races; a watermelon park; fireworks; and the giant fire pictures of Governor Sterling, Governor Murray, and the crossed flags of Texas and Oklahoma.

Durant was planning its own bridge-opening event. The editor of the *Durant Weekly* said he was authorized to announce that "the whole goggle-eyed world is invited and that the barbeque beef and pork will be free as the air we breathe! Concessionaires will be selling everything from ice cream cones to Budweiser, with plenty of free ice water and ample shade."[24]

Denison merchants were too busy at their cash registers to pay much attention to the latest National Guard deployment. Retailers saw increased sales resulting from the free bridge immediately. L. M. Newsome, owner of a popular soft goods store, was already making plans to expand his business. Mrs. R. E. Fehr made it a point to survey customers of her shoe store and found that a great number were from Oklahoma. Business at the Café DeLuxe was thriving, owing in large part to increased tourist traffic coming through town on the improved roads. On the first Saturday Market Day after the free bridge opened, thousands of shoppers milled through the business section, and plenty of Oklahoma license plates were in evidence. Managers of the Rialto Theater, the Safeway store, Burtis Drugs, Madden's Department Store, Busby's Jewelry, Jennings Furniture, and other retail businesses noted a lifting of spirits and willingness to spend that had been absent during the days the bridges were closed.[25]

Durant was seeing the benefits of increased traffic over the new bridge, too. During August, thirty young North Texas men and women applied for admission to Southeastern State Teachers College. Officials were confident that the new bridge would help boost attendance at that school. Durant also regained its wedding license crown. The county office issued 283 marriage licenses (mostly to Texans) in August, beating a monthly record set four years earlier.[26]

Governor Murray probably assumed his aggressive stance in order to draw attention from a worsening domestic situation in Oklahoma. Murray was forced to close the toll bridge over the Canadian River near Purcell when local farmers threatened to dismantle it and lynch the toll collector. (The governor sent highway department employees to scratch out a temporary detour around the toll bridge until a new free bridge could be constructed.) Unrest over the high price of ice led Murray to threaten the state's ice manufacturers with competition from ice manufactured and distributed by convict labor.[27]

By the end of the week, Murray reduced the force of Oklahoma National Guardsmen to nine men and two officers, and for two weeks the emplaced troopers watched traffic flow unhampered across the free bridge between Texas and Oklahoma.

Finally, on August 8, Murray withdrew the militiamen altogether. The Oklahoma governor chose to let stand the order establishing martial law at the crossings, however, declaring, "If they resume hostilities, we will reoccupy the military lines."[28]

Editorialists, columnists, and headline writers across the nation found in the armed conflict between two states an irresistible story. Before the arrival of the National Guard—even before the Texas Rangers rushed to the bridge to restore the barrier—dozens of newspapers around the country were describing to their readers news of the free bridge that federal courts shut down. But when Oklahoma highway workers forced open the bridge and legendary Texas Rangers closed it, editors moved the story to their front pages. The *Bakersfield Californian,* not known as a sensational newspaper, told its readers in an eight-column headline, "Oklahoma, Texas Near Civil War," whereas the *Syracuse* (NY) *Herald* promised readers the "Battle of the Red River" on its front page.[29]

News of the bridge war "is front page stuff in New York, Boston, Baltimore, and many of the leading dailies in that section," marveled one local editor. The *New York Times* and *Washington Post* provided daily coverage of Murray's deployment of the National Guard and of his visit to the battlefront.[30]

Few of the out-of-state newspapers reported on the seriousness or consequences of the conflict. Instead, the Red River bridge story served to take readers out of the darkness of other national news. President Hoover had recently announced a debt moratorium on reparations from starving European nations, former Harding cabinet member Albert Fall was on his way to prison for his involvement in the Teapot Dome scandal, police from all over the Midwest were tracking a gang of serial safecrackers who had broken into fourteen banks, and wildfires were tearing across much of Wyoming, Montana, and Idaho. The conflict over the Red River bridges displaced these more serious stories, turning the bridge war into a national event as outlandish as a dance marathon.

"Amusement is found by the greater part of the country in the battle which has been fought over bridges spanning the Red River," according to the *Newcastle* (PA) *News.* And humor there was. "Perhaps the marines should be sent to the Texas-Oklahoma border," the *Oakland* (CA) *Tribune* suggested. "A lesson in true neighborliness would seem to be in order." The *Oshkosh* (WI) *Daily Northwestern* acknowledged the "comic opera" aspects of the conflict and suggested that the governors "ought to hold a confab under a flag of truce and draw up a protocol to settle the dispute."[31]

Lighthearted as some of the coverage may have been, the stories introduced Alfalfa Bill Murray of Oklahoma as a politician willing to stand up for the needs of his people. As the bridge war drew to a close, Governor Murray was being lionized by the national press, characterized as something of a rustic folk hero, a man of action. Other elected officials—notably, President Hoover—might dither or order up a study; Alfalfa Bill was a man willing to pack his suitcase, arm up, and carry the fight to the front lines. World Wide News Service transmitted to its subscribers a close-up photo of the Oklahoma governor wearing a crisp campaign hat, gnawing a cigar, and gazing off into the distance as if daring any opponent to take a run at him. Many of the service's seven hundred newspapers—including the *Los Angeles Times* and the *Atlanta Journal-Constitution*—used the supplied caption: "None could honestly say he was not prepared for any eventuality." Other newspapers ran flattering feature articles. "Alfalfa Bill is ready to fight at the drop of a hat, Texans preferred," the *Rochester* (NY) *Times-Union* told its readers. The headline on a half-page profile in an eastern daily said, "Fiery governor of Western state carries out his own ideas and makes others like it," and an editorial cartoon in the *Kansas City Star* depicted Bill in knight's armor atop a powerful horse and brandishing a broadsword.[32]

Even in Texas, Murray's actions at the bridge were generally applauded by the press. The *Dallas Times-Herald* judged the Oklahoma governor worthy of "considerable acclaim as a man of courage."[33] The conservative *Dallas Morning News* conceded the fallacy of holding "that Governor Murray of Oklahoma is hare-brained" for his actions in the bridge matter. "He knows what he is about," the *News* said. "He is a man of cultivated brain and easy manner."[34]

Governor Sterling, however, took a drubbing. While it praised Murray as a man of courage, the *Dallas Times-Herald* wrote of Ross Sterling's timidity during the bridge standoff.[35] Likewise, the *Dallas Morning News* observed that Murray dealt, held, and played all the cards in the bridge conflict: "Murray handles this bridge play at both ends, just as if our Governor Sterling was dummy." Oklahoma newspapers were even more critical. "The Sterling administration has again crippled its usefulness by its attitude in the free bridge controversy," said the *Daily Ardmoreite*, and the plainspoken *Durant Weekly* editor said that the Texas governor's handling of the bridge matter "tended to show up Sterling for a boob—and a boob he is."[36]

During that hectic mid-July weekend in Houston when he signed the bill authorizing the bridge company to sue the state, Ross Sterling received a personal letter from C. G. Marshall, managing editor of *The United States Daily*, an influential pro-business journal in Washington, DC, that would eventually become *U.S. News & World Report.* Marshall asked Sterling for a profile of his career in business and politics, and a statement about the current bridge conflict. The Texas governor, who had never before dreamed of stepping onto the national political stage, responded immediately.[37]

On August 7, Sterling received another letter from editor Marshall, returning the governor's profile and bridge war statement. It didn't seem advisable for *The United States Daily* to publish the Texas governor's story at that time, Marshall said. "I'm quite sure you will appreciate our position under the circumstances." A week later, the influential journal published a profile of and statement by Governor Murray.[38]

By that date, the Red River Bridge War was over, and Alfalfa Bill Murray of Oklahoma was the clear winner.

CHAPTER 14

Put Away the Soldiers

AT THE HEIGHT OF the Red River Bridge War, the Nunnallee Furniture Company of Van Alstyne built a display in its Main Street show window. On a sand table, water ran through a narrow galvanized trough, representing the Red River. Miniature bridges crossed the stream and, on either end of the bridges, tiny autos lined up to cross. On the far side of the river were several toy soldiers, one with a machine gun pointed down the bridge, and an army truck. On the other side of the river were four tin cowboys, representing Texas Rangers. Mr. Rea Nunnallee placed a sign on his building announcing, "Conditions at Red River Today!" to draw stranded motorists off US Hwy. 75 for some impromptu shopping in his furniture store.[1]

On August 6, Mr. Nunnallee dismantled his model of the war zone, sweeping the last of the sand out of his store and giving the little toy soldiers and Texas Rangers to a neighborhood boy who delivered newspapers.

By August 8, 1931, most of the outstanding injunctions had been dissolved and peeled away like old wallpaper. No further hostilities showed themselves along the Red River, but most locals wondered at each crossing of the shining new bridge whether, on their return, they would see the open trusses of the $250,000 structure or a reconstruction of the barriers that would require another lengthy and tedious detour.[2]

The Oklahoma National Guard had withdrawn from the river and the rangers were scattered around Texas on other cases, but Little Joe Bailey and the Red River Bridge Company still had a war to fight. In October, Bailey filed suit against the Texas Highway Commission for $165,000, the value of the contracts signed by then-chairman Ross Sterling the previous year. The lawsuit was no surprise; it was specifically allowed by the state legislature to end the bridge standoff in July. Attorney General Allred, who would defend the suit, again maintained that the highway commission contracts were illegal, and he looked forward to showing that in court. Trial was scheduled in Austin after the first of the year.[3]

Bailey was confident he would prevail in his suit against Texas and its highway commission. It took him just two days in Austin to realize he was playing against a stacked deck.

In July 1930, before construction commenced on the free bridge that Alfalfa Bill would force open a year later, the president of the Red River Bridge Company signed contracts with the Texas Highway Commission. The contracts, signed by Ross S. Sterling as then-chairman of the commission, were intended to end all litigation and clear the way for completion of the new toll-free bridge over the Red River. The bridge company was giving up valuable assets: real estate, future toll revenues, and the exclusive toll-crossing franchise granted years before by the Chickasaw Nation. In return, the contracts called for the company to be compensated by Texas with a lump-sum payment and a monthly toll reimbursement if the free bridge opened early. Based on the contracts, the bridge company dropped all lawsuits, the free bridge opened ahead of schedule, and only then—a year later—did the Texas attorney general assert that the contracts were invalid.

The state had taken the bridge company's assets for its own with a contract it was now trying to disavow, as Joe Bailey saw it. Every day the newspapers were saying the same thing: "Owners of the toll bridges were possessed of some rights in the premises, else the federal court would not have recognized their plea for an injunction," said one newspaper.[4] "The bridge company owners should receive due consideration," wrote another.[5] Texans—most of whom were still in favor of hanging horse thieves—were big believers in private property rights, so surely a judge would force the highway commission to pay for the property it took.[6]

Bailey was wrong.

The case was assigned to the 126th District Court—a new division—in Travis County and would be presided over by W. F. Robertson, an Austin political insider who was recently appointed to the bench by Governor Sterling. (One of Robertson's early rulings prohibited Bailey from interrupting state business by requiring Governor Sterling and other state officials to submit to depositions before trial.)[7]

Trial opened January 11, 1932, with Bailey's motion that the state pay damages to the bridge company if the highway commission's contracts were proved invalid. The judge denied the motion, which left Bailey out in the cold if the judge agreed with Allred's position.[8]

Little Joe Bailey wasn't the bloviator in court that his father had been. His opening demurrer was a request that the judge assure compensation of the bridge company for its loss of the assets to the state. He read appropriate pages from the contracts, entered them into evidence, and then took his seat.

Attorney General Allred's opening statement was just as short and to the point: the contracts given to the Red River Bridge Company were invalid. Texas law—Jake Loy's original bridge bill—prohibited Texas from paying more than half the cost of the new free bridge. Besides, the contracts hadn't been signed by all members of the Texas Highway Commission and were not entered into the commission's official minutes.

Highway commissioner Cone Johnson was the first witness, and under oath he was much less specific about the bridge contracts than he had been in his statement to the *Dallas News* six months earlier. Shown the contracts, Johnson said he thought they were drafts, drawn up during negotiations with the bridge company. He recognized his signature, but didn't recall signing the documents.[9]

Judge W. R. Ely was the one of three highway commissioners who didn't sign the contracts. "I didn't want to pay anything," he said, "because the toll bridge operators were getting along pretty well already." Ely said the contracts were never discussed at a formal Texas Highway Commission meeting, and he had heard of them only when Senator Loy began his investigation. Asked why now-governor Sterling didn't bring up the matter at a commission meeting, Ely said that he and Cone Johnson were conducting most of the highway business then. "Governor Sterling was busily engaged in his campaign."

Gov. Ross Sterling took the stand Wednesday morning, responding genially to the questions asked him. He certainly signed the contracts, he said, but he signed without reading them during a hectic campaign stop in Gainesville on July 4, 1930. His approval of the contracts, he said, was strictly conditional on having the other two members of the commission sign and enter them into the minutes during a regular meeting. He had no idea that Ely had not approved the agreements, Sterling said, and he had no opinion as to whether the amounts promised in the contracts were appropriate or not. Throughout his testimony, Sterling gave the impression of a man uninvolved with the details of the bridge contracts.[10]

Allred rested his case.

Joe Bailey had only one rebuttal witness, a Dallas attorney named George H. Penland. A staunch Baptist, Penland admired Sterling's prohibition promises and his moral reputation, and was an early supporter of his campaign. Penland accompanied candidate Sterling on many of his North Texas campaign trips, including the one in Gainesville on July 4, 1930, where he was the one to hand Sterling the bridge contracts for a signature.[11]

Point by point, Penland refuted Sterling's testimony: Penland told Sterling that commissioner Ely wasn't going to go along with the contracts, and Sterling said Ely would go along eventually. Sterling spent up to twenty minutes reading the contract before signing it, Penland said, and the candidate said the bridge company should really have gotten a more equitable deal. Sterling said the contracts should be signed right away, "or the bridge people will stretch out litigation forever."

Bailey rested his case.

The next morning, Judge Robertson rendered his decision. "Bridge Company Loses Suit," was the eight-column headline on one newspaper.[12]

The judge had decided in lockstep with Allred's positon: the contracts were illegal and invalid; the state could not be made to pay. Additionally, the highway commissioners were simply careless or inattentive, the judge said. They could not be sued individually.

Bailey promised an appeal, but the state's involvement in further Red River bridge litigation essentially ended with Judge Robertson's verdict.

It is arguable as to whether Attorney General Allred knew that Governor Sterling's testimony in the Austin courtroom was false. If Allred didn't know of Sterling's falsehoods during the trial, he certainly learned soon after.[13]

Within moments of Judge Robertson's verdict against the Red River Bridge Company's contract (and while still in the same district courtroom), an obviously angry Allred announced he had just learned that the Red River Bridge Company payment contract was not the only one issued by the Texas Highway Commission in July 1930. Similar payment contracts were signed with the Gainesville-to-Ardmore and Terral-Ringgold toll bridge companies. Not only were there additional bridge payment contracts, but it was Allred's understanding that the Terral-Ringgold toll bridge owners had received a settlement payment from the Texas Highway Commission just four months earlier.[14]

Allred convened an immediate court of inquiry, commandeered a vacant jury room, and began issuing handwritten subpoenas for highway commissioners and Governor Sterling.

Only Cone Johnson, W. R. Ely, and Gibb Gilchrist, chief highway engineer, could be located and served that afternoon. Newsmen watched as Allred grilled the three men.

Yes, they said, they had drafted compensation contracts for all three bridge companies, believing the bridge owners deserved something for their assets and would tie up construction of the new bridges indefinitely if they didn't receive it. The amounts and terms of the contracts were decided at a closed conference prior to the start of construction on the bridges. No one remembered exactly how many, if any, of the negotiating conferences commission chairman Ross Sterling attended.

Johnson said that he and Sterling had signed all the contracts at the time, but commissioner Ely had balked at the amounts promised and refused to sign. Sterling had assured Johnson that Ely would eventually concur. When Allred declared the contracts invalid in July 1931, the Denison-Durant bridge owners had chosen to seek an injunction, but the Ringgold and Gainesville owners agreed to further negotiations through engineer Gibb Gilchrist.

According to Gilchrist, the Ringgold bridge owner settled for $15,000 the previous September. The group expected to settle with the Gainesville owner for a similar amount that very week. After eliciting promises that no further money would be paid without his approval, Allred adjourned his court of inquiry for the day.

The following morning, Allred announced he had located a treasurer's warrant issued September 16, 1931, authorizing payment of $15,000 to the Terral-Ringgold toll bridge company. The warrant had been requested by the three Texas highway commissioners and approved for payment by Governor Sterling.

At that point, Allred closed the hearing to newsmen.

There is no record to indicate that Allred continued this court of inquiry; there is no record to indicate what the results of the inquiry might have been.

Years later, confidential correspondence found in Governor Sterling's files reveal facts that Sterling did not volunteer and that he denied during the bridge contract trial.

On Monday, January 11, 1932—the day the bridge trial commenced in Judge Robertson's Austin courtroom—the governor received a personal letter from H. W. Stark of Gainesville. Stark owned a regional pharmaceutical distributorship and was a major shareholder in the company that owned the Gainesville-to-Ardmore toll bridge across the Red River. Stark had nominated Sterling for a seat on the Texas Christian University board of trustees and was a financial supporter of his gubernatorial campaign in North Texas.

"Dear Sir and Friend," Stark opened. "I do not want to bother you in regard to the Gainesville Red River bridge proposition with Judge Ely, but I do think that you should see him and talk the matter over with him, and see if we cannot get this business straightened out."

Stark was enlisting the governor's assistance with W. R. Ely (the highway commissioner who thought the bridge owners "were getting along pretty well already").

"I am sure that if you will talk to Judge Ely he will abide somewhat with your ideas in regard to the settlement. Although you are not a member of the commission now, this is part of your unfinished business."

Three days before Sterling would testify in court that he had no opinion as to whether the contract amounts were appropriate or not, Stark reminded him of an earlier conversation.

"You remember what you said to me in the Turner Hotel [in Gainesville] when you signed the contract. You said that you were sorry that you were not there, that we should have gotten a better deal out of it than we did get."

Stark closed his letter to the governor asking that he "do this much for me and insist on a settlement."[15]

Governor Sterling answered Stark's letter the same day. In his response he said he had spoken to Cone Johnson, Gibb Gilchrist, and (new commissioner) E. R. Martin. "They were open-minded," Sterling said, "and willing to do most anything. I can say further that Judge Ely will not object to adjusting your differences."[16]

Texas governor Ross S. Sterling lied—and probably perjured himself—about his actions and motives concerning the Red River bridge contracts. The Houston oilman with a reputation for never failing to honor his signature on a contract, the businessman known for getting deals done, took a shortcut to assure completion of the new Red River free bridges and then became entrapped in a tar baby of his own making.

No one revealed Governor Sterling's entire role in negotiating, drafting, signing, and settling the Red River bridge payment contracts. By the time of the Austin trial, Sterling was facing other problems.

After less than a year in office, it was obvious to most Texans that Sterling was the wrong man at the wrong time to deal with the state's deficits, joblessness, oil and cotton problems, and a recalcitrant legislature. The *New York Times*, in a major piece on Sterling, observed that "the public sees in him a distraught executive, the too-frequent spectacle of the successful businessman learning that government is not business." The article commented on the number of supporters who were deserting him and on the likelihood he would be a one-term governor. "Governor Sterling has tried to occupy an untenable position in public office," the *Times* continued, "that of trying to please everybody."[17]

Ross Sterling tried to please everybody in the Red River bridge conflict: Texans who wanted free crossing of the border river, bridge owners who wanted fair compensation for their property, and the laws of the state he was sworn to serve. Sterling failed them all.

He was the fourth casualty of the Red River Bridge War.

When Joe Bailey and the Red River Bridge Company lost their legal battle in Texas, they took it to Oklahoma.

Bailey still had his action pending in the Muskogee federal courtroom; Judge Colin Neblett had chosen to step away from the bridge conflict, so the next hearing was held in the Kansas City courtroom of John C. Pollock, a seventy-four-year-old jurist who brooked no lack of respect in his court. Joe Bailey renewed his claim against Oklahoma for forcibly opening the new free bridge and for denied access to the toll bridge, rendering the toll bridge essentially worthless. He built the case against Oklahoma for disregarding the original Chickasaw Nation by building the free bridge closer than two and a half miles from the toll bridge, costing the bridge company the value of its franchise.[18]

Little Joe added another action: a provision of the original payment contract with Texas required the bridge company turn over to Oklahoma three acres of property on the north bank of the river, property on which Oklahoma built the north end of the bridge. Now, because Texas had ruled the contracts invalid, Bailey wanted Oklahoma to remove the north end of the bridge and return the property.

The Red River Bridge Company was asking Judge Pollock to award them a quarter of a million dollars (or the right to operate the free bridge as a toll bridge, collecting seventy-five cents from every vehicle that crossed over it until the quarter of a million dollars was collected).[19]

W. C. Lewis appeared in the Kansas City courtroom in June 1932, arguing that Judge Pollock should dismiss the suit entirely. A private entity may not sue a state without that state's permission, he argued. The judge refused to dismiss the action. "Either your state owes money to the bridge company or it doesn't," he told Lewis.

Pollock told both attorneys to return the following week to make their arguments.

Alfalfa Bill was furious. The US Constitution couldn't be any clearer, he stormed to reporters. He quoted Article III, Section 2: "In all cases in which a state shall be a party, the Supreme Court shall have original jurisdiction."

"The free bridge will be kept open until the Supreme Court of the United States decides otherwise, Judge Pollock notwithstanding," the governor asserted. Murray reminded reporters that his declaration of martial law along the Red River was still in place. That morning, he said, he had once again deployed his militia to the Red River, and men were once again guarding the north and south ends of the free bridge.[20]

"I don't care what they do about their toll bridge," Murray said, "but I am concerned with keeping that free bridge open."

Federal judge John C. Pollock had been assigned to the federal bench thirty years earlier by Pres. Theodore Roosevelt; before that he had been a justice of the Kansas Supreme Court. He was not about to take a scolding or constitutional law lessons from anyone, and certainly not the governor of Oklahoma. At seventy-six years old he was, in many ways, as opinionated and outspoken as Alfalfa Bill. "[Judge Pollock] is the quintessence of reaction, vested rights, and federal power," one attorney warned Murray. "The old man is just too stubborn to back down."[21]

Pollock refused to dismiss Bailey's suit against Oklahoma, finding that the Chickasaw Nation franchise was valid and binding. After two days of argument, the judge sent his opinion to all parties on October 26, 1932. "The franchise conferred a property right on Colbert and his successors," Pollock wrote, "that even the government itself cannot take for public purposes without just compensation being made."[22]

Bailey had waited four years to hear those words. The Chickasaw franchise had been exactly what Big Joe Bailey was prepared to argue when he dropped dead in the Sherman courtroom. Pollock's next words were even more welcome.

"I think," Judge Pollock said, "that Oklahoma is entitled to a damage trial before a jury."

A jury trial was the last thing Oklahoma's legal team wanted. A jury trial meant that twelve ordinary people would determine the bridge company's damages, and the state would have no choice but to pay the bill. It was a twelve-way crapshoot; anything could happen.

The judge's decision sent Oklahoma's attorney general J. Berry King scrambling for a strategy. Allred and the state of Texas had no interest in seeing the free bridge closed, so he and Benbow spent several days in Oklahoma City with King and Lewis devising a course of action. Trial was scheduled for April 1933.[23]

In the end, the only course of action that mattered was Governor Murray's, and his strategy was simple: boycott Pollock's courtroom.

"You are hereby notified not to appear in that case further," Murray wrote his attorney general, "because the state has no interest in the case. Since the state has not given permission for any [citizen or corporation] to sue the state on the controversy involved, and since no inferior federal court has jurisdiction to try this state," the trial is of no concern to Oklahoma.[24]

Murray went on to say that Judge Pollock was also wrong to find that the Chickasaw franchise still existed. Though the Chickasaw Nation may have been a sovereign one, its sovereignty was delegated by Congress and Congress, in the act that created the state of Oklahoma, expressly prohibited "any perpetuities or monopolies."

"You are therefore instructed," Murray repeated, "and also all of your deputies, not to appear in said case for any purpose whatsoever."

Attorney General King was stupefied. This was a case that could cost Oklahoma taxpayers a quarter of a million dollars or more, and Alfalfa Bill opted out! The governor is given wide latitude under Oklahoma's free-wheeling state constitution, King knew, but this made no sense. Regardless, King bit his tongue and shut up about it for the time being.

Murray *was* concerned about one part of the suit. The three acres of land under the north end of the bridge had been deeded to the state and, based on a county appraisal, Oklahoma had paid the bridge company $88 without having gone through a formal condemnation process. There was a chance, Murray knew, that a jury could decide that a value of tens of thousands of dollars might be fairer.

Murray hired Malcolm Rosser, a journeyman attorney from Muskogee, to represent the state on the right-of-way matter, and Rosser added his voice to the chorus warning Murray about Judge Pollock. "He considers Oklahoma as a sort of outlaw state," Rosser wrote the governor, "and will try very hard to find a way to render a large verdict against it."[25]

Murray remained unperturbed.

Little Joe Bailey, no stranger to complex and lengthy trials, arrived in Judge Pollock's federal courtroom on Monday morning, April 10, 1933, for what must have been one of the strangest days of his legal experience. With no one to object, Bailey seated his jury, made his opening statement, and presented his case, all before breaking for lunch. Rosser presented no witnesses, but entered into evidence the paperwork showing that Oklahoma had paid the bridge company for the three acres of right-of-way.[26]

"By the close of evidence," Rosser wrote, "the judge seemed to listen with impatience and to have his mind fully made up to require a heavy judgment against us." The only hopeful sign, Rosser thought, was that the Judge instructed the jury to determine the value of the land, the bridge, and the franchise separately. Separate values would make a final settlement easier to negotiate.

"The jury seemed to have some trouble in arriving at a verdict," according to Rosser, "but just before adjournment this afternoon they returned."

"*Red River Company Is Given $168,000 Damages*"[27]

It could have been worse. The jury set the value of the Chickasaw franchise at $100,000. ("It is easily worth a half a million," Rosser wrote Governor Murray.) The bridge was valued at $67,000. ("As junk, it is too high.") The three acres of right-of-way had a value of just a thousand dollars. ("If we only have to pay that, I shall not feel that any injustice was done," Rosser said.)[28]

Rosser thought Judge Pollock could be convinced to reconsider his decision on the validity of the Chickasaw franchise. "Unless the old man is just too stubborn to back down, I feel sure we can convince him that he cannot enter judgment" on the franchise and bridge amounts, Rosser advised the governor.

Alfalfa Bill wasn't in the mood to convince anyone. Asked about the verdict and likely judgment against the state, Murray unloaded on Judge Pollock, calling him a criminal and an insurrectionist for violating his oath to the Constitution.

"He had no more jurisdiction to do that than had a justice of the peace in some remote precinct of Arkansas," Murray said. "We'll pay the thousand dollars, but we won't pay a damn cent more."[29]

Any chance of negotiation evaporated in the heat of the governor's words. Judge Pollock entered a judgment for $168,000 against Oklahoma, warning that he would allow the toll bridge company to take over the free bridge if Oklahoma refused to pay the amount due.[30]

"That free bridge is going to be open as long as I'm governor if it takes five thousand National Guardsmen to keep it open," Murray shot back.

"May I suggest that any federal judge who commits such acts had better get on an airplane and get out of the state to avoid arrest."

Despite strong public criticism by Attorney General King and members of the legislature for a flawed strategy costing Oklahomans $168,000, Alfalfa Bill stood firm in this new bridge conflict. He ordered King to appeal Pollock's judgment all the way to the US Supreme Court, if necessary.[31]

On June 18, 1934, the US Court of Appeals for the Tenth Circuit reversed Judge Pollock on exactly the grounds Murray predicted. The Chickasaw franchise was not valid; Oklahoma could not be held party to a damage suit brought in an inferior federal court. Oklahoma's only obligation was a thousand dollars for three acres of right-of-way property.[32]

Hearing the news, a smug Alfalfa Bill Murray lit another cigar and grinned. "Under the United States Constitution, no other decision could have been rendered," he said.

In January 1936, the Oklahoma highway department paid $1,211 to the Red River Bridge Company in compensation for the small plot of land on the north bank of the Red River that was now under the north entrance to the toll-free highway bridge. Oklahoma's role in the Red River Bridge War of 1931 officially ended.[33]

Later that year Jake Loy and a group of Denison business owners urged the state of Texas to purchase and maintain the old toll bridge, feeling it would help the city open new trade areas. During its 1937 session, the Texas legislature appropriated $50,000 to purchase the bridge, approaches, and any franchises.[34]

In 1938, a Texas Highway Department clerk added the toll bridge—built over the Red River in 1915 to aid automobile travel between Denison and Durant at the site of Frank Colbert's ferry crossing—to a list of 25,000 bridges in the state's bridge inventory, where it was soon forgotten.[35]

EPILOGUE

Historical Markers

SHORTLY AFTER his toll bridge judgment was overturned on appeal, Judge Pollock was playing cards with fellow Kansan Joseph Taggart, who had served in the US Congress with Governor Murray.

"How are you and my friend Alfalfa Bill getting along?" Taggart teased the judge.

"Didn't he hand me a load?" Pollock laughed. "And the worst part of it is, he was right."[1]

To one degree or another, the Red River Bridge War marked every individual connected with the conflict. Some used it to improve their careers; others used it to find new careers. Most never forgot it; at least one was never forgotten for it.

Joseph W. Bailey Jr. continued his work on behalf of the Red River Bridge Company even as he ran for and won election to the US House of Representatives in 1932. Little Joe left the House for an unsuccessful Senate run against a popular incumbent and returned to his Dallas practice full-time. He always regretted that he didn't win his legendary father's final case. In 1942, at the age of fifty, he was commissioned as an officer in the US Marine Corps. He suffered fatal injuries in an auto crash and died near Gainesville, just a stone's throw from the Red River, in 1943.[2]

Jake Loy resigned his Texas senate seat in 1932 to run for Grayson County judge, an office he held for the next thirteen years. Loy remained involved in county road issues but made his mark as a proponent of juvenile justice issues. His belief in rehabilitation through recreation led him to oversee development of a public recreation area and lake outside Denison. Loy Park and Loy Lake are named after him. Jake Loy died in 1945.[3]

Reporter Claud Easterly spent his entire forty-seven-year newspaper career with the *Denison Herald*, interviewing five US presidents, Father Flanagan,

Joe Louis, and dozens more luminaries of the twentieth century. His wife, the woman he referred to as his "War Bride," never held him at fault for having cut their honeymoon short. He retired in 1972 as managing editor and vice president of the company. On retirement, he took with him the manual typewriter he used to type his Red River Bridge War stories and continued using it for the next twenty-five years until he passed away in 1999.[4]

Burt McCaleb was a district engineer in Bryan County when he and his group of workers sparked the Red River Bridge War by bullying their way past a hapless watchman and dismantling the bridge barricade. He spent the rest of his career with the Oklahoma Highway Department, much of it as a design engineer. His son, Neal, served as Oklahoma secretary of transportation.

A rare (for the time) Texas Republican, Judge Thomas M. Kennerly remained on the federal bench in Houston, deciding some significant civil rights and suffrage cases. Judge Colin Neblett, who threatened to arrest Governor Murray for contempt, returned home to New Mexico. He retired from the bench in 1948 and served as chairman of the state fish and game commission for twenty years. Acerbic Judge Pollock of Kansas City died of pneumonia in 1937.[5]

Adjutant General W. W. Sterling and Texas Ranger Tom Hickman were effectively fired (along with all other active rangers) during a political purge in 1933. Sterling rose to the rank of colonel during World War II and, after the war, managed the Driscoll ranches in South Texas until he passed away in 1960. Hickman became a sheriff's deputy in Gainesville, taking off from time to time to participate in rodeos. The reorganized Texas Rangers were merged with the Texas Department of Public Safety in 1935. Hickman was appointed commissioner, then chairman, of the Public Safety Commission. He died in 1962 at seventy-six years old.[6]

Col. John MacDonald was often called upon as an after-dinner speaker to talk about his role in the Red River Bridge War. He maintained his law practice in Durant and kept his seat in the state senate while remaining active in the National Guard. At age forty-six he was stricken with a heart attack and died in 1939.[7]

Assistant Rice Tilley left the Texas attorney general's office at the end of 1930 and was sworn in as special legislative counsel to the state senate. He helped the senate draft legislation on vehicle registration, fuel taxes, and regulation of billboards next to state highways. Tilley later moved to Fort Worth and entered private practice.[8]

Sidney Benbow left Austin in 1934 and returned to Houston as a regional director of the new Federal Housing Authority. In 1942, he returned to his family farm and took up a life of farming, ranching, and the occasional practice of law.[9]

After wrapping up the bridge case, Oklahoma assistant attorney general W. C. Lewis was named US district attorney for Oklahoma. Lewis and family eventually moved to Washington, DC, where he made the ritual motion before the US Supreme Court to admit his wife and son to practice there.[10]

Texas attorney general James Allred ran a successful gubernatorial campaign in 1934. His campaign materials listed his accomplishments as attorney general, and never failed to note the $165,000 he saved taxpayers in Texas by defeating the claims of the Red River Bridge Company.[11]

The Red River Bridge War marked no one more than the governors of Texas and Oklahoma. Ross Sterling and William H. Murray shared the public eye that summer of July 1931, their methods and manners compared and contrasted. The harsh spotlight helped send one into political obscurity and the other onto the pathway to the presidency.

Ross S. Sterling was turned out of office at the end of his first term. The bridge war—which occurred six months into his term—was not the only reason for his defeat, but it portrayed Sterling as something akin to the officious postal clerk or the stingy banker. Texas' millionaire governor was considered no friend of the common man when he denied use of a new free bridge to the people who struggled to pay the toll taker. "Governor Sterling's position is comparable to the local conception of President Hoover," said the *New York Times*. "He may be a good man, but not for public office during these times."[12]

"Few appreciated the crushing stress I was under in the latter part of 1931 and 1932," Sterling wrote years later. By the end of his term Sterling's fortune—oil and real estate holdings, investments, company shares, interlocking loans, everything—was gone, eaten by the Depression.[13]

Sterling refused to attend the inauguration of his successor and left Austin for Houston as soon as possible. Using his wife's hoarded hundred-dollar Liberty bond as seed money to raise investment capital, he began to rebuild his personal wealth. At sixty years old he was still a man willing to hit four licks while other men hit just two or three, and by the time he passed away in 1949, he was once again rich (but not oil rich).[14]

Despite the shortcuts he took with the bridge contracts and his lack of candor after, there is absolutely no evidence that Ross Sterling was a venal politician. Quite the opposite. His life—before and after his term as governor—is marked throughout by incidents of charity and kindness. He is most often described as a likeable man of no great political skill who encountered extreme economic conditions he could not control. Few fail to mention his revitalization of the Texas Highway Department. More than any other man in the twentieth century, Ross Sterling brought Texans out of the mud and onto the highways.[15]

There was a brief moment, some say, when Gov. William "Alfalfa Bill" Murray held the 1932 Democratic Party presidential nomination in his grasp.

Murray first earned national headlines with the sortie into Texas to open the free bridge, but his invasion of that state was just beginning; immediately upon ending the conflict at the Red River, Murray embarked on a victory tour. Invited to the Old Settlers' Reunion in Weatherford on August 14, Alfalfa Bill drew a crowd of ten thousand farmers and families eager to hear the hero of the bridge war. The following day, a crowd of fifteen thousand people gathered in Dallas for a speech by the Oklahoma governor. "He is a master of the appeal to the masses," the *Dallas News* marveled. Some writers speculated that Governor Murray was just the man to run against Herbert Hoover in the upcoming presidential elections.[16]

During the Bridge War people across the country couldn't seem to hear enough about Alfalfa Bill. And the more they heard, the more they wanted to hear. During fall 1931, Murray sat for adoring interviews with *Time, The Literary Digest, Collier's, Saturday Evening Post, American Mercury, Commonweal,* and other prestigious national magazines.[17]

"Murray's popularity," according to *Collier's*, "is largely founded on the popular belief that Alfalfa Bill is the champion of those opposed to the rich." In 1931, the people of a still largely rural America appreciated a man who knew firsthand how falling crop prices could cost a family its land, why the rising price of bread was not just an economic problem but a moral outrage, and why those two things happening together meant that the banks and the corporations were faring much too well during hard times.

Murray remained coy about his intentions but scheduled speeches before huge crowds in Nebraska, Colorado, Wyoming, the Dakotas, Kansas, and other farming states, staging them with the requisite bands, bunting, and buttons carrying the campaign slogan of "Bread, Butter, Bacon, and Beans." Many of those states were committed to their own favorite sons, but Murray's supporters said privately that the delegations would turn to Alfalfa Bill on later ballots.[18]

The governor formally announced his candidacy in February 1932 with words that alienated party politicians of the smoke-filled rooms. He promised that he "would make no trades, form no combination no compromise on any principal for delegates or for the nomination." Asked whether he would support the party platform, Alfalfa Bill said flatly, "I'll say what I please."[19]

Murray went to the national party convention in Chicago as Oklahoma's favorite son, and a political alliance with four other states' rights populists might have allowed Murray to match Franklin D. Roosevelt and Al Smith vote for vote. But Murray's inflexibility, some political gaffes, and a side deal cut by Texan John N. Garner left the Oklahoman out of the running after the first ballot.

"My administration has just been one damn thing after another," he told a reporter.[20] Unable to succeed himself, Murray left the governor's mansion in 1935 and retired to a small farm with no electricity or running water.

Murray was never again elected to public office, but he made a half-hearted run in the gubernatorial primary of 1938. Alfalfa Bill's denunciations of Roosevelt, the New Deal, and his personal eccentricities cost him the nomination. His wife died the same year, which accelerated the mental and physical decline of the sixty-nine-year-old polymath.

In 1951, he administered the oath of office to his son, Johnston Murray, the fourteenth governor of Oklahoma. He lived for a while in the mansion with his son and daughter-in-law, picking up pocket change by selling copies of his memoirs in the state capitol building. Few people recognized the skinny, bent old man carrying the bag of books.

Alfalfa Bill's last days were spent in Tishomingo. He was a shambling, unkempt figure who was always good for an outrageous quote to the newsmen who occasionally sought him out. One of those newsmen was novelist Tony Hillerman, working at the time for United Press in Oklahoma. "I am left with one of those intensely graphic memories," Hillerman wrote, "of a proud old man in a dingy room who shares a bathroom down a long hallway with the same poor folks he tried to help."[21]

Alfalfa Bill Murray died in Tishomingo in 1956. He is buried there, in a cemetery adjacent to the college that bears his name.

"The toll bridges have outlived their usefulness," the *Dallas Morning News* pronounced in 1932. "Once that free bridge was opened it couldn't have been closed by the President himself," another newspaper echoed.[22]

The Red River Bridge War marked the end of the private toll bridge era in twentieth century America. By the end of Alfalfa Bill Murray's administration there was no toll bridge operating on any intrastate or interstate thoroughfare in Oklahoma. Texas still had the international toll bridges in partnership with Mexico, but the "Pay Toll Here" sign had disappeared from Texas highways.[23]

Though it was welcomed as a regional resource when it was built in 1915, the Red River automobile toll bridge was, by 1930, an economic burden on Depression-stricken residents and a detriment to regional development. In just fifteen years—from 1915 to 1930—the Denison-Durant private toll bridge (and most others across the country) fell victim to technological changes, a depressed economy, higher costs of capital, and the changed expectations of motorists and transportation planners. By the end of the twentieth century, the privately owned toll bridge was virtually extinct in the United States.

Often with much heated discussion, states and state transportation agencies began a return to tolled turnpikes and bridges in the latter part of the century. European nations, however adopted another model: the public-private partnership (PPP). Under a PPP, the jurisdictional entity—a city, a regional authority, a state—grants a private company an exclusive concession to build and operate a toll bridge or road. The jurisdictional entity receives payment based (usually) on some combination of up-front payment, periodic payments, an override on toll collections, and a per-vehicle amount.[24]

The popularity of PPPs in Europe, Africa, the Middle East, Australia, and Asia—projects totaling more than $600 billion since 1985—led American states to consider long-term infrastructure concessions to private companies. A 2013 review by the National Conference of State Legislatures counted forty-two states with transportation PPP-enabling legislation. Scores of private concession projects have been financed to date, including four large projects in Texas.[25]

As tempting as public-private transportation projects may be to cash-strapped states and municipalities, events at the Red River bridges in 1931 provide some object lessons for citizens of the twenty-first century.

The Red River toll bridge operated under long-term franchises granted by the US Congress, the Texas legislature, and the Chickasaw Nation. As did Frank Colbert, private transportation concessionaires require lengthy agreements to assure a return on construction expenses while reassuring investors and lenders. But change can occur quickly.

The pace of technological change in just the past two decades has been dizzying, but so it was a hundred years ago as the United States transitioned from horse and buggy to motorized personal transportation. The entire nature of roads and bridges—placement, construction materials, specifications, maintenance, and a hundred other details—had to be reinvented and implemented, then rereinvented and done over again each year as vehicles improved. The Red River Bridge Company's toll bridge was perfectly engineered for the automobile of 1913 when the bridge opened in 1915. The 1913 Ford Model T runabout weighed twelve hundred pounds and had a safe cruising speed of forty miles per hour; the 1930 Ford Model A weighed twice as much and cruised at sixty miles per hour. Just fifteen years after it opened, the Red River toll bridge was approaching obsolescence: narrow and unsuited for heavy two-way traffic at speed. Fast-changing automotive technologies could leave the public saddled with an outmoded, inefficient transportation network.

Red River Bridge Company shareholders were probably pleased with the possibility of selling their franchise and bridge to Texas and Oklahoma when the two states announced plans to build a new one. The alternative for the company was to build an updated new bridge of their own. Texas and Oklahoma shared the eventual $300,000 cost of the toll-free bridge in 1931 and financed it with bonds issued by the states. For the toll bridge company to replace their bridge with a new structure, however, would have required it to raise the entire amount on its own. The cost of capital—if a lender could be found in the nervous investment environment of 1931—would have been much greater than the cost of tax-free government highway bonds.

If Oklahoma, Texas, and the Red River Bridge Company had entered into a PPP—and *if* the bridge company could have financed a new bridge and paid up-front concession fees—how would the states have enforced state-of-the-art safety upgrade and maintenance requirements on the bridge company? What if the bridge company couldn't afford the upgrades or maintenance as they were needed? Private capital will always cost more than public monies, and may not always be available when the private partner needs it.

The concessionaire engaged in a PPP is driven by a profit motive, but profit-driven transportation planning and operation do not always bring about the best solution for affected communities. The Red River Bridge Company (and Frank Colbert himself in the earlier years) operated a tolled river crossing that provided a necessary service for the communities near the river. But, as economic depression began to smother trade in the late 1920s, it became onerous to nearby communities and thwarted efforts of state governments to spur commerce and recovery. A state agency might find it easier to suspend tolling in the case of a natural disaster or evacuation emergency, for example, but a private partner might be less willing to cooperate with that public need.

An increasing number of American bridges and roads will again become pay-for-use in the twenty-first century, and some will be built, operated, and maintained through PPP agreements. That may be good business for the private companies, but maybe not so good for the communities involved. Citizens groups, municipalities, state agencies, and traffic planners will do well to learn the lessons of the toll bridge era of the early twentieth century and of the Red River bridge conflict in the Depression summer of 1931.[26]

Every day more than thirty thousand vehicles cross the Red River, driving north and south on US Hwy. 75 between Denison and Durant. Some will exit the highway at the first off-ramp south of the Red River—Texas exit #75—to visit the Texas Travel Information Center. There, drivers and passengers can browse the travel brochures, visit the restrooms, or just stretch their legs by walking the landscaped grounds.[27]

A few curious visitors might notice a metal plaque mounted at eye level on a pole near the big metal star. The twenty-seven-by-forty-two-inch plaque is a standard Texas Historical Commission marker, this one describing the "1931 Free Bridge Controversy." Though the visitor couldn't know it from the two-hundred-fifty-word text, a hundred yards north of where he is standing is where Texas Rangers W. W. Sterling and Tom Hickman built the barricade that halted traffic over the free bridge in that summer of 1931. Then, if the visitor should turn around and look directly south down the road toward Denison, he would notice the slight rise over which the US Hwy. 75 Association caravan arrived—horns blaring, lights shining, and band playing—to be among the first to cross the new bridge when it opened on July 25, 1931.

It is impossible to see the old toll bridge from the marker. However, a half-mile walk due east through the underbrush would lead to the bank of the Red River, just about where Frank Colbert landed his ferries more than a century ago.

The Texas Travel Information Center on US Hwy. 75 north of Denison is as good a place as any to mark the geography of the Red River Bridge War, though there's not much left to see.

By 1951, the silvery new free bridge had tarnished to a muddy gray. Oklahoma and Texas grudgingly cooperated to contract for a new paint job, and the bridge sparkled again. This stretch of the King of Trails Highway, the much-vaunted Gulf Coast-to-Canada route of the early auto age, was bypassed for interstate highway status in 1952. Even so, increasing traffic required Texas and Oklahoma in 1958 to cooperate once again to build a companion to the free bridge, running parallel and adjacent to one another. The two bridges—one for northbound traffic, the other for southbound—carried up to twenty thousand vehicles a day across the Red River for most of the rest of the twentieth century.[28]

The old toll bridge at Colbert's Ferry outlasted both the governors who fought over it, but by 1960 it was in sad shape. The bridge flooring had rotted through in places, wooden side rails were dropping into the river, and the steel stringers were rusted through in places. With toll collectors long gone, locals and bootleggers still used the old span as a shortcut across the river. A Grayson County official wanted to sell the bridge to a scrapper, but could find no records of current ownership. After a three-month search through Texas Department of Highways bridge records, clerks finally located the toll bridge hiding in plain sight, and it was deeded over to Grayson County.[29]

Before the county could deal with the old toll bridge, a gas line running under it ruptured in 1961, and the resulting fire burned the center span. Ironworks in Sherman and Durant bid to remove the rest of the bridge for salvage, leaving only tall stone pilings projecting from the bed of the river.[30]

Bridge engineering had made great strides in the sixty years since the free bridge had opened in 1931. Modern beam bridges could incorporate more safety features and were easier to maintain than the old steel truss bridges, so Texas and Oklahoma again cooperated to build a new bridge across the Red River. They scheduled the old 1931 free bridge for demolition.[31]

At 1:30 p.m. on December 6, 1995, engineers used 750 pounds of explosives strapped on the stone piers to drop the center truss of the bridge into the Red River. Traffic would be temporarily rerouted pending completion of the new bridge. A Texas Department of Public Safety spokesperson said the rest of the bridge would be cut up and removed, but an Oklahoma woman from Colbert lobbied successfully to preserve part of the historic span.

Some motorists driving north or south on US Hwy. 75 between Denison and Durant might glance at a large park on the east side of the highway about two miles north of the river. They may see an eighty-foot-long section of steel highway bridge mounted in the middle of the green space—old-fashioned steel chords, struts, and bracing forming a lattice on each side of the span—and wonder how the bridge came to be there.

If they were curious enough to pull off the highway and stop, they might see families from Colbert and nearby communities at the park bridge enjoying picnics, ball games, or concerts. In the evening, they may notice sweethearts strolling across it under moonlight.

And if they were to ask a local how this old bridge came to be set down in the middle of a park in Colbert, Oklahoma, they might be told it's a souvenir of a day when the toll taker collected seventy-five cents from every automobile, bus, truck, and wagon to cross the thousand feet over the Red River.

They might be told that this section of the 1931 free bridge is a reminder of the Red River Bridge War and of a time when a free bridge could make the difference between starvation and survival. They might learn how a new bridge could bring hope to the families nearby.

ACKNOWLEDGMENTS

IF YOU READ this book, enjoyed it, and maybe learned something from it, you should be aware that a lot of good people helped make it possible, and I'm proud to acknowledge their contribution.

One of the first of those is Wyota Hannan of Colbert, Oklahoma. I met her at a potluck picnic dinner in Colbert City Park, in the shadow of a section of the original 1931 free bridge, where she had organized a community lecture about the bridge war. Later, Wyota brought together a group of longtime residents of the Colbert and Denison areas willing to talk to me about their early days by the Red River. Charlene Barnett, Laverne Sims Bledsoe, Faye Carroll, Mickey Fisher, Mary Catherine Hodge, Donna Hunt, Mike Mather, Marilyn Miller, and Charles Scott spent a frigid February day in the Colbert Methodist Church meeting room recalling hard times, loving families, the character of the Red River, and the blessings of a free bridge. James Hannan described a long-ago Sunday walk to the river where he, his brother, and his father met Alfalfa Bill.

Ralph Elder provided valuable research assistance in Austin, as did Tonia Wood in the archives division of the Texas State Library and Archives Commission. Additional research assistance came from John Kemp of the Wallisville Heritage Park museum and archives; Bill Welge, Terry Zinn, and John May of the Oklahoma Historical Society; Heather Bateman and Ashley Hosek of the Carl Albert Center; and Russell Perkins of the Oklahoma Department of Transportation. Individuals at Dallas Public Library, Metropolitan Library System (Oklahoma City), Reynolds Library (Durant), Sherman Public Library, Denison Public Library, and Frontier Village (Denison) were unfailingly helpful and encouraging.

Rice M. Tilley Jr. was kind enough to share his father's papers relating to the bridge war. Thanks also to members of the McCaleb, Benbow, Ashburn, Hughes, and Rosser families, who related family stories, confirmed dates, and shared documents.

Durant attorney Ken Rainbolt guided me through legal issues related to tribal law, Oklahoma-Texas border disputes, navigable waterways, eminent domain, and the difference between injunctions and restraining orders. If errors of legal interpretation remain, they are entirely my responsibility.

Dr. and Mrs. Chris Gallup provided the perfect writer's retreat—the privacy of their snowed-in cabin in the Colorado Rockies.

Seven people did me the great favor of plowing through an early manuscript of this book, writing detailed critiques of clarity, readability, character, and much more: Daryl Ayalla, Ken Freehill, Kyla Harden, Crystal Kreitzer, Laura O'Kelley, Ken Rainbolt, and Suzi Thornton.

The staff of Texas A&M University Press was always professional and supportive.

Throughout this book project (and others) I've benefitted from the enthusiasm and support of my son, Julian. Meagan Alma and Ava Kai were always on my mind. Even as I first learned to hold a book, my mother, Pat Williams, was sharing with me the joy and power of the written word. Now a nonagenarian, she is still my biggest supporter and fan.

NOTES

Introduction

1. Description of C. T. Davis from US Census, 1920 and 1930, Grayson County, TX.

2. This account of the confrontation between Davis and McCaleb is from "Hickman and Sterling on Job at Span," *Denison Herald*, July 17, 1931.

3. "Motorists Have Inning as Free Span Is Temporarily Opened," *Denison Herald*, July 17, 1931.

4. "Texas Governor Orders Closing," *Sherman Democrat*, July 17, 1931, "Highways to Toll Bridges Blocked by Murray's Order," *Daily Oklahoman*, July 17, 1931.

5. Emily Thornton, "Roads to Riches," *Business Week*, May 7, 2007, and Jeffrey Spivak, "Toll Roads: The Route to Redevelopment?" *Urban Land: The Magazine of the Urban Land Institute*, October 29, 2013.

Chapter 1

1. Charles A. Cerami, *Jefferson's Greatest Gamble* (Naperville, IL: Source, 2004).

2. James E. Lewis, *John Quincy Adams: Policymaker for the Union* (Wilmington, DE: Rowman & Littlefield, 2001).

3. The text of the Adams-Onís Treaty of 1819 (rendered in English) is available at http://www.tamu.edu/faculty/ccbn/dewitt/adamonis.htm

4. Completion of Denison Dam and impoundment of Lake Texoma in 1944 reduced flooding along the upper Red River, but the river can still strike without warning.

5. Harry Sinclair Drago, *Red River Valley: The Mainstream of Frontier History from the Louisiana Bayous to the Texas Panhandle* (New York: C. N. Potter, 1962), and Arthur C. Benke and Colbert E. Cushing, *Rivers of North America* (Amsterdam: Elsevier/Academic, 2005).

6. Grant Foreman, "Red River and the Spanish Boundary in the Supreme Court," *Chronicles of Oklahoma* 2 (March 1924): 298–310

7. Kerry C. Kelly, "Maps of Indian Territory, the Dawes Act, and Will Rogers' Enrollment Case File." National Archives and Records Administration. See www.archives.gov/education/lessons/fed-indian-policy/

8. Arrell M. Gibson, *The Chickasaws* (Norman, OK: University of Oklahoma Press, 1971) and Michael W. Lovegrove, *A Nation in Transition: Douglas Henry Johnston and the Chickasaws, 1898–1939* (Ada, OK: Chickasaw, 2009).

9. Grant Foreman, *Down the Texas Road: Historic Places along Highway 69 through Oklahoma* (Norman: University of Oklahoma Press, 1936).

10. Waterman Ormsby, correspondent for the *New York Herald*, was one of the first through passengers on the Butterfield Overland Stage, and some of his dispatches to the *Herald* are reprinted in Grant Foreman, "The California Overland Mail Route through Oklahoma," *Chronicles of Oklahoma* 9, no. 3 (September 1931): 312. For the newsman's description of the entire journey, see Waterman Lilly Ormsby and Lyle Henry Wright, *The Butterfield Overland Mail* (San Marino, CA: Huntington Library, 1942).

11. Benjamin Franklin Colbert most often called himself "B. F. Colbert" in business records and filings, but family members, employees, and newspaper writers during his lifetime referred to him as "Frank." (See "Mortuary: B. F. Colbert," *Dallas Morning News*, March 14, 1893.) By the mid-1930s, some writers were incorrectly referring to him as "Ben Colbert." (See "Home of Early Colbert Ferryman," *Sherman Democrat*, July 26, 1931.) For a concise history of the Colbert family, see G. B. Braden, "The Colberts and the Chickasaw Nation," *Tennessee Historical Quarterly* 17 (1958): 222.

12. This physical description of Frank Colbert comes from a photograph accompanying W. B. Morrison (recorder), "Colbert Ferry on Red River, Chickasaw Nation, Indian Territory: Recollections of John Malcolm, Pioneer Ferryman," *Chronicles of Oklahoma* 16, no. 3 (September 1938): 305, and from a description given by his son in "One Time Indian Envoy at Capital," *Sherman Democrat*, July 26, 1931.

13. Foreman, "The California Overland Mail Route," 313.

14. Most of the firsthand descriptions of Colbert's Ferry and its operations in this chapter come from Morrison, "Colbert Ferry on Red River," 302–13. John Malcolm was born in Scotland on July 2, 1845, and died on April 6, 1934. He arrived on the Red River in 1870 and for most of the next thirty years worked with Frank Colbert as a ferry operator, engineer, and business partner. Two additional interviews inform this chapter. Interviews with S. T. Atkins (12595) and Lula Austin (7334), both conducted in 1937, describe Colbert's Ferry operations. The interviews can be found in the Indian Pioneer Collection, part of the Western History Collections at the University of Oklahoma at Norman.

15. "State Items," *Dallas Herald*, November 30, 1872.

16. The Chickasaw franchise is incorporated into the record of the Circuit Court of Appeals, Tenth Circuit, State of Oklahoma ex. King, Attorney General, v. Handy, June 18, 1934. This is an appeal from the District Court of the United States for the Eastern District of Oklahoma; John C. Pollock, Judge.

17. Foreman, "The California Overland Mail Route," 312.

18. "Indian Territory," *New York Times*, May 15, 1887, describes Frank Colbert as "among the richest" with a worth of approximately $150,000 to $200,000. The small town of Colbert, Oklahoma, still thrives. Ask residents how the town's name is pronounced, and they will tell you it is spoken as "call-bert," not "cole-bert," and certainly not "cole-bear."

19. The Missouri, Kansas, and Texas Railroad was the first national line to link with the spreading web of Texas railroads, and the connection would be made at Denison. Walter R. Borneman, *Rival Rails: The Race to Build America's Greatest Transcontinental Railroad* (New York: Random House, 2010).

20. Jack Maguire, *Katy's Baby: The Story of Denison, Texas* (Austin: Nortex, 1991).

21. Morrison, "Colbert Ferry on Red River," 311.

22. See *Congressional Record*, 43rd Cong., 1st sess., vol. 2, 66–68, for letters of recommendation and transmittal.

23. Morrison, "Colbert Ferry on Red River," 311.

24. Hans Peter Mareus Neilsen Gammel, *Laws of Texas 1822–1897* (Austin: Gammel Book Publishing, 1898), 7:52–54.

25. See *Congressional Record*, 43rd Cong., 1st sess., vol. 2, for text of the bill.

26. "Denison News," *Dallas Weekly Herald*, June 6, 1874.

27. Morrison, "Colbert Ferry on Red River," 311.

28. "Denison 62–50–35 Years Ago," *Denison Press*, August 28, 1937, reprinting an article originally appearing August 28, 1875.

29. Morrison, "Colbert Ferry on Red River," 312.

30. Ibid.

31. Gammel, *Laws of Texas*, 82–84. Corporate officers and major shareholders are listed in a Denison news wrap-up, "Held Up by Highwaymen," *Dallas Morning News*, May 10, 1893. They include a railroad executive, a surgeon, a factory president, and a bank owner as well as Frank Colbert and his son.

32. "Big Bridge Completed," *Dallas Morning News*, October 31, 1891, and "Denison Doings," *Dallas Morning News*, November 17, 1891.

33. "Roads for Denison," *Dallas Morning News*, February 2, 1905.

34. "Contract for $50,000 Bridge," *Dallas Morning News*, January 26, 1913.

Chapter 2

1. An interview with Green and his description of the wild ride can be found at "Mr. E. H. R. Green's Carriage," *Dallas Morning News*, October 6, 1899.

2. Green remained an enthusiastic motorist. He was an organizer and first president of the Dallas Automobile Club and board chairman of the Texas State Automobile Association. For more biographical information, see "First Auto Seen in Dallas Causes Fright, Amazement," *Dallas Morning News*, October 7, 1951, or "The Witch, the Wench, and the Colonel," *New York Press*, February 14, 2009.

3. For autos produced in 1904: *Census of Manufactures: 1905* (Washington, DC: US Government Printing Office, 1906), 272. Average cost: *Official Good Roads Yearbook of the United States* (Washington, DC: American Highway Association, 1914), 1910, 135. Autos in Dallas and Tulsa: "Automobiles," *Dallas Morning News*, January 1, 1905; "Highways," in *Encyclopedia of Oklahoma History and Culture* (Oklahoma City: Oklahoma Historical Society) http://www.okhistory.org /publications/encyclopediaonline

4. Descriptions and specifications of autos through 1931 come from Floyd Clymer, *Those Wonderful Old Automobiles* (New York: McGraw-Hill, 1953).

5. D. C. Greer, *History of the Texas Highway Department*, rev. ed. (Austin: Texas Highway Dept., Information & Statistics Division, 1959), and Annual Report of the State Highway Commission, 1924. *Biennial Report of Oklahoma Department of Highways*. Oklahoma City: Oklahoma Dept. of Highways. 1914–1932. Also see "Hastening the Millennium," *Saturday Evening Post*, March 4, 1911, 26–29.

6. As a sponsor of the endurance race, the *Daily Oklahoman* gave daily coverage to the event. See "Big List for Endurance Run," *Daily Oklahoman*, May 18, 1910, and "Sixteen Autos Start on Run," *Daily Oklahoman*, May 30, 1910.

7. "Racers End State Dash," *Daily Oklahoman*, June 5, 1910.

8. Greer, *History of the Texas Highway Department*, and viewing archival photos at Oklahoma Department of Transportation, Oklahoma City.

9. Jeffrey Brown, "The Numbers Game: The Politics of the Federal Surface Transportation Program" (PhD diss., University of California, 2003), 144–48.

10. John E. Potter, *United States Postal Service: An American History* (Washington, DC: Government Relations, USPS, 2007), 22.

11. Earl Swift's *The Big Roads: The Untold Story of the Engineers, Visionaries, and Trailblazers Who Created the American Superhighways* (Boston: Houghton Mifflin Harcourt, 2011) is a remarkably readable history of the US scramble to provide improved roads for an ever-increasing number of automobiles. Karl Edward Wallace's "Texas and the Good Roads Movement: 1895 to 1948 (master's thesis, University of Texas at Arlington, 2008) describes the physical, financial, and political challenges of creating highways in Texas and the Southwest. The *Official Good Roads Yearbooks*, published from 1901 to 1921, are interesting because of their boosterism and engineering detail.

12. Bob Burke, *ODOT 100: Celebrating the First 100 Years of Transportation in Oklahoma* (Oklahoma City: Oklahoma Heritage Association, 2011), 4.

13. "Betterment of Roads," *Dallas Morning News*, February 9, 1905.

14. "Roads for Denison," *Dallas Morning News*, February 1, 1905.

15. See *Yearbook of the United States Department of Agriculture*, for appropriate years.

16. Greer, *History of the Texas Highway Department*.

17. Annual Report of the State Highway Commission, 1919. *Biennial Report*

of Oklahoma Department of Highways. Oklahoma City: Oklahoma Dept. of Highways. 1914–1932.

18. Jack Maguire, *Katy's Baby: The Story of Denison, Texas* (Austin: Nortex, 1991).

19. "Will Build Red River Bridge," *Dallas Morning News,* December 14, 1912, and "Contract for $50,000 Bridge," *Dallas Morning News,* January 26, 1913.

20. "Red River's Large New Bridge Open," *Dallas Morning News,* March 28, 1915.

21. "Revival of Turnpike Companies," *Southern Good Roads,* January 1910, 8.

22. "Red River Bridge Finished," *Dallas Morning News,* February 27, 1915, and "Bridge Over Red River Is Finished," Dallas Morning News, March 12, 1915.

23. "Value of More Good Highways," *Dallas Morning News,* July 14, 1912.

24. "Crop Prices Aided by Good Roads," *Daily Oklahoman,* July 25, 1913. Also see "Auto Trucks Cut Marketing Costs," *Daily Oklahoman,* September 28, 1913, and Isaac B. Potter, *The Gospel of Good Roads: A Letter to the American Farmer,* 2d ed. (New York: League of American Wheelmen, 1891).

25. "Meridian Highway Becoming a Fact," *Dallas Morning News,* March 28, 1915, and "Many Points of Interest on Line of New Highway," *Daily Oklahoman,* April 12, 1914.

26. "Trails Manager Talks to Club," *Tulsa World,* May 17, 1922.

27. US House of Representatives, *Red River Bridges. Hearing before the Committee on Interstate and Foreign Commerce, February 10 and 11, 1930.* 71st Cong. (Washington, DC: Government Printing Office, 1930).

Chapter 3

1. Karl Edward Wallace, "Texas and the Good Roads Movement: 1895 to 1948 (master's thesis, University of Texas at Arlington, 2008), 63–64. Wallace writes of county inefficiencies and waste during the years 1915 to 1925. As early as 1911, however, Robert J. Potts of the pioneering highway engineering school at Texas A&M University was warning of the potential for large-scale graft inherent in county control of roads. See the text of his speech to the Texas Conservation Congress at "Conservation of Material for Good Road Building Important," *Fort Worth Star-Telegram,* February 12, 1911.

2. Annual Report of the State Highway Commission, 1916. *Biennial Report of Oklahoma Department of Highways.* Oklahoma City: Oklahoma Dept. of Highways. 1914–1932.

3. D. C. Greer, *History of the Texas Highway Department,* rev. ed. (Austin: Texas Highway Dept., Information & Statistics Division, 1959), and Wallace, *Texas and the Good Roads Movement.*

4. "Injunction Sought in Harris Court," *Fort Worth Star-Telegram,* August 18, 1925.

5. Lenora Nickels, "Public Services of Dan Moody" (master's thesis, Texas Tech University, 1948).

6. Dan Moody's legal victory against Klan violence is described in "Jury at Georgetown Returns Verdict in Burleson Flogging Case," *Dallas Morning News*, September 26, 1923, and it cemented his reputation for moral rectitude. He was elected attorney general and served in one of the most corrupt administrations in Texas history. The administration was that of Miriam "Ma" Ferguson, wife of former governor James "Farmer Jim" Ferguson. Jim was impeached for corruption in 1917 and prohibited from running for governor again. In 1924, Jim ran his wife for the office, publically stating that he would be calling the shots around the capitol. While Mrs. Ferguson presided over state business, Jim was selling highway contracts, pardons, and everything but the silverware in the governor's mansion. For more on the Ferguson administrations, see Carol O. Wilson, *In the Governor's Shadow: The True Story of Ma and Pa Ferguson* (Denton: University of North Texas Press, 2014), or Norman D. Brown, *Hood, Bonnet, and Little Brown Jug: Texas Politics, 1921–1928* (College Station: Texas A&M University Press, 1984).

7. "Jury Holds Case Should Be Heard in Travis County," *Fort Worth Star-Telegram*, February 7, 1926.

8. "Big Men Needed on Highway Board," *Fort Worth Star-Telegram*, November 24, 1925. According to a story in the same issue, "Virtually every newspaper in Texas was demanding [the] summary dismissal" of Texas highway commission chairman Frank V. Lanham and Commissioner Joe Burkett. See "Lanham Summoned as He and Burkett Resign," *Fort Worth Star-Telegram*, November 24, 1925.

9. The biographical information about Ross S. Sterling in this chapter comes primarily from Ross S. Sterling and Edward W. Kilman, *Ross Sterling, Texan: A Memoir by the Founder of Humble Oil and Refining Company* (Austin: University of Texas Press, 2007). This memoir originated as a series of lengthy interviews of Sterling by Houston journalist Ed Kilman in the 1940s, which Kilman converted to a standard third-person biographical narrative. When Kilman's manuscript failed to interest a publisher, the manuscript was set aside for sixty years. Don Carleton, executive director of the University of Texas at Austin's Dolph Briscoe Center for American History, edited and revised Kilman's manuscript, returning it to a first-person narrative. Nowhere in this detailed memoir does Sterling mention events connected to the Red River Bridge War. See also Henrietta M. Larson and Kenneth Wiggins Porter, *History of Humble Oil & Refining Company: A Study in Industrial Growth* (New York: Harper, 1959); Ross Phares, *The Governors of Texas* (Gretna, LA: Pelican, 1998); and "Ross Sterling Looms Big in Texas Affairs," *Dallas Morning News*, July 11, 1930.

10. Sterling and Kilman, *Ross Sterling, Texan*, 8.

11. Sterling and Kilman, *Ross Sterling, Texan*, 13.

12. A joke of the time had two friends meeting on the street in Houston. "Have you been to Beaumont lately?" one man asks the other. "Yes, I was there day before

yesterday," was the reply. Shaking his head, the first man says, "You wouldn't know the place now!"

13. Sterling and Kilman, *Ross Sterling, Texan*, 26.

14. Sterling and Kilman, *Ross Sterling, Texan*, 23.

15. Sterling and Kilman, *Ross Sterling, Texan*, 24.

16. Larson and Porter, *History of Humble Oil & Refining Company.*

17. Sterling and Kilman, *Ross Sterling, Texan*, 41.

18. The physical description of Sterling from this time comes from a lengthy profile of his background and business philosophy: "Ross Sterling Looms Big in Texas Affairs," *Dallas Morning News*, July 11, 1930.

19. Sterling and Kilman, *Ross Sterling, Texan*, 60.

20. Betty T. Chapman, "Diverse Legacy of Ross Sterling Still Evident All Over Houston," *Houston Business Journal*, September 14, 2001.

21. Sterling and Kilman, *Ross Sterling, Texan*, 80.

22. "Big Men Needed on Highway Board," *Fort Worth Star-Telegram*, November 24, 1925.

23. Sterling and Kilman, *Ross Sterling, Texan*, 80.

24. Sterling describes his strategy for preempting opposition by the oil industry in Sterling and Kilman, *Ross Sterling, Texan*, 83. Also, "Ross Sterling Looms Big in Texas Affairs."

25. The *Fort Worth Star-Telegram*, for example, ran a lengthy Roads Report for its West Texas readership every Thursday for more than a decade.

26. The newspaper editor devoted two of his This Evenin' columns to reader complaints about toll bridges. This motorist comment is from "Toll Bridges Are in Bad with Most Folks," *Daily Ardmoreite*, February 20, 1928. For more public opinion, see "Toll Bridges," *Daily Ardmoreite*, February 15, 1928.

27. "Toll Bridge System in State Is Criticized by Editor of Texas Paper," *Daily Oklahoman*, October 30, 1927.

28. "Free Bridges Favored by Denison Merchants," *Dallas Morning News*, September 23, 1923.

29. The charges were borne by the Eufala Commercial Club and announced in their ad. *Eufala Indian Journal*, October 27, 1927.

30. "Jefferson Highway Popular," *Dallas Morning News*, October 15, 1921. The Jefferson Highway was, like the King of Trails, a named (and heavily promoted) highway route. The Jefferson Highway linked New Orleans and Winnipeg, and crossed the Red River at Denison.

31. "Nobody Loves a Toll Bridge," *Daily Oklahoman*, December 2, 1928.

32. "Easy Money in Toll Roads," *Port Arthur News*, September 20, 1927.

33. "Toll Highways Menace US Motor Owner," *Dallas Morning News*, May 27, 1928, and "Automobilists Hit Toll Road System," *New York Times*, June 29, 1928.

34. "New Trinity River Bridge in Prospect," *Dallas Morning News*, August 26, 1927, and "Toll Bridge across Red River Considered," *Dallas Morning News*, September 25, 1927.

35. "The Flea Aspires to Be a Big Toll Bridge and Ferry Man from Oklahoma," *Daily Ardmoreite*, March 6, 1928. The *Daily Oklahoman* repeated the *Tulsa World* survey later that year and asked legislators for public comment. See "Toll Bridges Are Sounding Their Death Knell," *Daily Oklahoman*, September 16, 1928.

36. "Open Free Pease River Bridge to Public," *Dallas Morning News*, October 12, 1926. The seventeen toll bridges over the Red River are cataloged in "Toll Bridges Are Sounding Their Death Knell," *Daily Oklahoman*, September 16, 1928.

37. Biographical information comes from Donna Hunt, "Judge Jake Loy," *Texoma Living!*, no. 3 (June 1, 2007); the Texas Legislative Reference Library; and "Death Takes Judge Loy at Sherman," *Dallas Morning News*, July 5, 1945. The federal spending limit is described in "Federal Aid Refused on Hwy. 91," *Denton Record-Chronicle*, February 21, 1928.

38. "Bridge Aid Measure Approved by Senate," *Dallas Morning News*, March 16, 1927, and "Representative Loy and House Clerk Wed," *Dallas Morning News*, June 3, 1927.

39. "Moody Approves Eight Measures," *Dallas Morning News*, March 30, 1927.

40. "Oklahoma to Aid Texas on Bridges," *Fort Worth Star-Telegram*, April 9, 1927.

41. "To Expend Big Sums for New Highways," *Dallas Morning News*, March 28, 1927, and "Federal Road Aid Soon Will Be Available," *Dallas Morning News*, March 31, 1927.

42. The maverick car dealer was Hal Collins, "Interesting Bits about Our Friends," *Denison Press*, May 5, 1941; Devol arson: "Failed Once, Fire Bug Burns Bridge," *Fort Worth Star-Telegram*, May 18, 1927; Gainesville arson: "Gainesville Bridge Burns," *Dallas Morning News*, December 1, 1927.

43. "Red River Toll Span Hearing in Sherman," *Fort Worth Star-Telegram*, April 12, 1928.

44. "Hearing Is Had on Red River Bridge Bill," *Dallas Morning News*, February 8, 1927, and "Mayfield Bill Is Authorized," *Dallas Morning News*, March 5, 1927.

45. A concise (and understandable) explanation of navigable law for non-lawyers is provided by Harold Kelso, "'Navigable Waters' as a Legal Fiction," *Journal of Land & Public Utility Economics* 17, no. 4 (November 1941): 394–405.

46. "Texas Senator Reviews Work," *Dallas Morning News*, April 13, 1927.

47. "Candidates for Representative," *Ada Evening News*, May 20, 1928.

48. "Conference Called over Bridge Tolls," *Dallas Morning News*, January 7, 1928.

49. "Bridge Toll Appraisal Committee Will Meet," *Dallas Morning News*, March 20, 1928.

Chapter 4

1. "Toll Bridges," *Washington Post*, March 13, 1929. There has been little scholarly attention paid to the rise of anti-toll sentiment in the 1920s, but newspaper editorials, letters to the editor, public statements by lawmakers, and the actions of state and national legislators bear out that opinions ran hot on the issue.

2. "House Decides to Retain its Special Rules," *Daily Oklahoman*, February 21, 1929, and "Bridge Bill's Death Is Seen," *Dallas Morning News*, February 23, 1929. Coke Johnson: "$532,678 Road Contracts Awarded," *Fort Worth Star-Telegram*, January 22, 1929.

3. The public hearings are described in "Toll Bridge Rate Fought," *Dallas Morning News*, June 5, 1928; "Pro Rata Toll on Bridges Up," *Dallas Morning News*, June 6, 1928; "Auditor Tells Operation Cost of Toll Bridges," *Dallas Morning News*, June 7, 1928; and "Bridge Hearing Ends; 2 Sides Pat on Tolls," *Fort Worth Star-Telegram*, June 9, 1928. The War Department order is at "Fix Rates for Toll Bridges," *Dallas Morning News*, January 6, 1929.

4. "Toll Bridges Are Discussed," *Dallas Morning News*, February 9, 1929, and "Bridge Bill's Death Is Seen," *Dallas Morning News*, February 23, 1929.

5. "Red River Toll Reduction Suit Is Considered," *Dallas Morning News*, April 6, 1929, and "State Contends Bridge Franchise Is Long Expired, *Wichita Daily Times*, April 7, 1929.

6. "Red River Bridge Suit Expected to Be Heard in Sherman on Friday," *Sherman Democrat*, and "State Petition Filed against Bridge Tolls," *Fort Worth Star-Telegram*, both April 10, 1929.

7. "Bridge Is Again Enjoined," *Fort Worth Star-Telegram*, April 12, 1929.

8. An account of the brief bridge opening and moving the toll booth, as well as the text of Judge Newman's order, can be found in "Bridge Owners Evade Oklahoma Injunction," *Sherman Democrat*, April 12, 1929.

9. My account of the events in Sherman on April 13, 1929, comes primarily from four newspapers, each of which had a reporter present on that day: the *Denison Herald*, *Sherman Democrat*, *Dallas Morning News*, and *Fort Worth Star-Telegram*. See their issues of April 14, 1929, and April 15, 1929. In addition, I referred to a series of articles written by Judge R. C. Vaughn that appeared in his Other Voices column in the *Herald Democrat* (of Sherman and Denison) on July 18, 2004, August 1, 2004, August 8, 2004, August 15, 2004, August 22, 2004, August 29, 2004, September 5, 2004, and September 12, 2004. Judge Vaughn based his columns, in part, on recollections tape-recorded by attorney Rice Tilley in the 1970s, and an interview with Sherman furniture retailer Forest Peveto, conducted and recorded in 1979. In the cases of conflict between the accounts, I relied on the earliest.

10. Biographical information on Joseph W. Bailey comes from "Bailey, Joseph Weldon—Biographical Information," US Senate Historical Office. See also Ben R.

Guttery, *Representing Texas: A Comprehensive History of US and Confederate Senators and Representatives from Texas* (Charleston, SC: BookSurge, 2008); Sam Hanna Acheson, *Joe Bailey, the Last Democrat* (New York: Macmillan, 1932); and Ray E. Karrar, "Invention in the Congressional and Campaign Speaking of Joe Weldon Bailey" (master's thesis, North Texas State University, 1968).

11. From Robert A. Caro, *The Years of Lyndon Johnson*, vol. 1, *The Path to Power* (New York: Vintage, 1990), 308.

12. Biographical information for Joe Bailey Jr. comes from Guttery, *Representing Texas*, and "Joseph Weldon Bailey, Jr. '15," *Princeton Alumni Weekly*, October 15, 1943, 9. The McKinley story is from Acheson, *Joe Bailey*.

13. It is difficult at this distance in time to discern Bailey's precise legal strategy. The Dallas Historical Society holds files from his private legal practice through 1920. Presumably, the Red River Bridge Company's legal files and Bailey's client correspondence and personal notes remained with Little Joe after his father's death. After Junior's death, his practice was closed and files put in storage. I've been unable to determine whether those files still exist after eighty years.

14. "Bridge Company Strikes Back in Toll Suit," *Sherman Democrat*, April 22, 1929.

15. "State Asks Return of Toll Bridge Case," *Daily Ardmoreite*, April 24, 1929.

16. "Tolls Now Collected by Receivers Named by Federal Court," *Sherman Democrat*, April 28, 1929.

17. "Injunction against Bridge Company Granted," *Sherman Democrat*, May 3, 1929.

18. "New Bridge Suit Is Filed by State," *Dallas Morning News*, May 2, 1929.

19. "Last-Minute Rush Marks End of Special Session," *Daily Oklahoman*, July 6, 1929, and "Toll Bridge Measure Wins," *Daily Ardmoreite*, July 11, 1929.

20. "Plan That Tolls Pay for Bridge," *Dallas Morning News*, October 23, 1929.

21. "Toll Bridge Plans under Discussion," *Dallas Morning News*, November 9, 1929, and "Bridge Contracts over Red River to Be Let Nov. 25," *Dallas Morning News*, November 9, 1929.

22. Nowhere in his memoir or in press accounts of his public comments does Sterling speak to the issue of toll bridges in Texas. I found only one instance—an off-the-record response to the publisher of the *Sherman Democrat*—in which Sterling offers his opinion on toll bridge issues. "I feel that the [bridge] owners should be paid by the respective states a fair replacement value of said bridges" (Sterling to J. Newell Johnston, December 24, 1929, R. S. Sterling Collection, Wallisville Heritage Park, Research Center & Library, Wallisville, TX). His only *public* words on the matter would be taken under oath eighteen months later.

23. For a recap of the accomplishments of the Texas Highway Department under Sterling's stewardship, see "R. S. Sterling, Millionaire, to Run for Governor," *Dallas Morning News*, May 30, 1930, and "Department Won Stature by Evolution," *Dallas*

Morning News, September 16, 1963. Sterling wasn't shy about trumpeting those accomplishments in his campaign ads. For example, see *Fort Worth Star-Telegram*, July 6, 1930.

24. Cotton prices quoted throughout come from auction prices for "middling" grade in Fort Worth, as published daily in the *Fort Worth Star-Telegram* through the 1920s and 1930s. Prices for other grades at other cities might vary by a few pennies, but all cotton prices rose and (mainly) fell in lockstep across North Texas and southern Oklahoma.

25. Ross S. Sterling and Edward W. Kilman, *Ross Sterling, Texan: A Memoir by the Founder of Humble Oil and Refining Company* (Austin: University of Texas Press, 2007), 92.

26. "Judge Urges Parley on River Bridge Proposal Squabble," *Wichita Daily Times*, June 5, 1930.

27. "Chickasaw Act Held Invalid," *Daily Ardmoreite*, March 21, 1930, and "Chickasaw Toll Bridge Act Minus Authority," *Dallas Morning News*, March 22, 1930.

28. "Red River Bridge Conference Held," *Dallas Morning News*, May 22, 1930.

29. "Bridge Company Asks Injunction," *Daily Ardmoreite*, June 11, 1930.

30. Sterling and Kilman, *Ross Sterling, Texan*, 95.

31. Aubrey: "Another Texas Bank Closes," *Big Spring Herald*, February 7, 1930; Post: "J. M. Pope Trial Continues," *Lubbock Avalanche*, March 26, 1930; Ennis: "First National Bank in Ennis Fails to Open," *Fort Worth Star-Telegram*, February 11, 1930; Merkel: "Merkel Bank Closes," *Wichita Daily Times*, February 10, 1930; Jefferson: "Jefferson Bank Closes," *Dallas Morning News*, February 11, 1930; Henderson: "Henderson Bank Closes," *Dallas Morning News*, February 25, 1930; Brownwood: "Brownwood Bank Closes," *Dallas Morning News*, June 27, 1930.

32. Several (slightly) different versions exist concerning how Sterling came by his "Fat Boy" nickname. See "Rises from Poor Farmer Lad to Highest Office," *Mexia Weekly Herald*, January 23, 1931, and "'Ma' Ferguson Will Face Second Vote Test," *New York Times*, July 26, 1932.

33. Sterling and Kilman, *Ross Sterling, Texan*, 108, 121.

34. It is apparent that reporters were doing more speculating than reporting on these closed-door meetings: "Conference Is Called on Red River Bridge," *Fort Worth Star-Telegram*, June 24, 1930. They did note the Oklahoma delegation's abrupt departure: "Bridge Settlement Still Uncompleted," *Dallas Morning News*, June 25, 1930.

35. Sterling's only public discussion of the contract would occur in an Austin courtroom eighteen months later. See the account of Sterling's sworn testimony about these contracts in chapter 14.

36. Sidney Benbow to Hon. R. S. Sterling, November 19, 1931, Records of Ross S. Sterling, Texas Office of the Governor, Archives and Information Services Division, Texas State Library and Archives Commission.

37. "Texas Asks Old Toll Bridge Rates," *Daily Ardmoreite*, August 3, 1930.

38. "Work Being Pushed on Red River Bridge," *Dallas Morning News*, November 17, 1930.

Chapter 5

1. "'Alfalfa' Bill Sells His Eden in Bolivia and Plans Return to Oklahoma," *Daily Oklahoman*, July 5, 1929.

2. "Bolivian Colonists Back Disillusioned," *New York Times*, April 6, 1925, and "Bolivia Colony Chief to Come Home," *Los Angeles Times*, July 12, 1925.

3. "Nick Longworth Now Headed for the Senate Say Political Prophets," *Daily Ardmoreite*, March 13, 1925.

4. William H. Murray, *Memoirs of Governor Murray and True History of Oklahoma: Together with His Biography, Philosophy, Statesmanship, and Oklahoma History, Interwoven.* (Boston: Meador, 1945), 2:263.

5. "Two Tulsans to Bolivia," *Tulsa World*, December 28, 1919.

6. "Bolivia Colony Chief Returns," *Los Angeles Times*, July 12, 1925.

7. "Nick Longworth Now Headed for the Senate Say Political Prophets," *Daily Ardmoreite*, March 13, 1925.

8. Murray, *Memoirs*, 2:265.

9. The biographical material about William H. Murray in this chapter was derived from three main sources. Keith L. Bryant's *Alfalfa Bill Murray* (Norman: University of Oklahoma Press, 1968) is by far the most factual, and it was my primary resource. Gordon Hines's *Alfalfa Bill: An Intimate Biography* (Oklahoma City: University of Oklahoma Press, 1932) was written as a campaign biography in 1932. It is generally accurate on biographical details and contains some period anecdotes about Murray. Murray's *Memoirs* is the least accurate, although it was written by Murray himself. It is as much lecture as autobiography, and his unique voice comes through the pages as he retells each life story to his benefit or to settle scores.

10. Bryant, *Alfalfa Bill Murray*, 13.

11. Ibid., 22.

12. Ibid., 23.

13. My primary sources for early Oklahoma history were Edward Everett Dale and Morris L. Wardell, *History of Oklahoma* (New York: Prentice-Hall, 1948); Edwin C. McReynolds, *Oklahoma: A History of the Sooner State* (Norman: University of Oklahoma Press, 1977); and Michael W. Lovegrove, *A Nation in Transition: Douglas Henry Johnston and the Chickasaws, 1898–1939* (Ada, OK: Chickasaw, 2009).

14. The elegant Chickasaw National Capitol Building still dominates downtown Tishomingo. An adjoining period building contains a museum of tribal history and culture.

15. "Gov. Johnston's Ambition," *Daily Ardmoreite*, September 7, 1903.

16. Bryant, *Alfalfa Bill Murray*, 24.

17. There are several versions of the nickname story, but they differ only slightly. I've used the version from Hines, *Alfalfa Bill*, 145.

18. Bryant, *Alfalfa Bill Murray*, 29.

19. Johnson, *Brief Biography*, 21.

20. "Hon. W. H. Murray Plans Vigorous Campaign," *Muskogee Times Democrat*, September 25, 1905.

21. For more information about this rare statehood convention for a territory that never became a state, see Bryant, *Alfalfa Bill Murray*, 36–39.

22. "Murray the Choice for President of Convention," *Ada Evening News*, November 20, 1906.

23. "Alfalfa Bill Murray, Hanratty the Winners," *Muskogee Times-Democrat*, November 20, 1906.

24. "Alfalfa Bill," *Daily Ardmoreite*, November 30, 1906. The eventual selection of mistletoe was an emotional one. During the previous decade, winters had been exceptionally hard ones in the Territory. Settlers had died of the cold and ice, and there were no flowers to put on their graves. The only thing with a bit of color in that brutal winter landscape was mistletoe. Mistletoe was named the state flower—commemorating those early settlers—and remained so until 2004, when it was replaced by the Oklahoma Rose. *The Oklahoman*, June 11, 2012,

25. Bryant, *Alfalfa Bill Murray*, 83.

26. Ibid., 120.

27. "Bits," *Sallisaw Star-Gazette*, August 14, 1914.

28. His Chautauqua program was reprinted in the *Daily Ardmoreite*, March 14, 1918.

29. "Alfalfa Bill," *Daily Ardmoreite*, October 20, 1915.

30. "Bill Balks at Begonias," *Muskogee Times-Democrat*, September 11, 1914.

31. Reinhard H. Luthin, *American Demagogues: Twentieth Century* (Gloucester, MA: Peter Smith, 1959), 108.

32. Murray, *Memoirs*, 234.

33. "Tishomingo Folk Prepare to Welcome Return of Wm. Murray," *Daily Ardmoreite*, August 22, 1929.

34. Hines, *Alfalfa Bill*, 268.

35. This story is part of the Alfalfa Bill legend, and each version has its own minor variations and embellishments. I've used Louise Overacker and Victor J. West, *Money in Elections* (New York: Macmillan, 1932), 90.

36. Murray, *Memoirs*, 2: 388.

37. "The Return of 'Alfalfa Bill' Murray," *Chronicles* 41, no. 1 (1963): 38–65.

38. "Rapid Progress on Bridge near Denison," *Dallas Morning News*, January 12, 1930.

Chapter 6

1. "Americans in Depression and War," in *The US Department of Labor: Bicentennial History of the American Worker* ed. Richard B. Morris (Washington, DC: US Department of Labor, 1976), 26–31.

2. Ross S. Sterling and Edward W. Kilman, *Ross Sterling, Texan: A Memoir by the Founder of Humble Oil and Refining Company* (Austin: University of Texas Press, 2007), 128.

3. "Jobless Estimated at 50,000 in State," *Miami News Record*, December 8, 1930, and "Census Report as to Jobless Is Incomplete," *Dallas Morning News*, September 15, 1930.

4. "Tomorrow Is Election Day," *Miami News Record*, April 6, 1931.

5. "Farmers Riot in Arkansas, Seeking Food," *Daily Oklahoman*, January 4, 1931; "Victims of Drought Threaten Food Riot," *Washington Post*, January 4, 1931; and Dale Ingram and Max Brantley, "The Forgotten Rebellion," *Arkansas Times*, January 19, 2006.

6. "7 Arkansas Towns Have Breadlines," *New York Times*, January 9, 1931. During the earliest years of the Great Depression, food riots received national press attention. Later in the decade, they became so common as to be all but ignored. (See Irving Bernstein, *The Lean Years: A History of the American Worker* [Baltimore: Penguin, 1970].)

7. "Hoover for US Aid as Very Last Resort," *Fort Worth Star-Telegram*, February 4, 1931.

8. On its editorial page the *Daily Oklahoman* listed the institutions that would not provide assistance. "The government as a unit of relief, is out of it," topped the list. "The one method left to the people is the relief of the needy with money donated by the public and administered by the Red Cross." See "The Chosen Way," *Daily Oklahoman*, January 19, 1931.

9. Arkansas: "7 Arkansas Towns Have Breadlines," *New York Times*, January 9, 1931. Clarksville: *Dallas Morning News*, January 21, 1931. Soup: "Soup and Coffee on Nickel Ticket at Parish House," January 2, 1931.

10. "Sun Beams on Ross Sterling as He Takes Governor's Oath," *Dallas Morning News*, January 21, 1931. Fort Worth: Sterling and Kilman, *Ross Sterling, Texan*, 180.

11. All quotes in this paragraph from "Murray Takes Over Governor's Office, Greets His Friends," *Daily Oklahoman*, January 13, 1931.

12. "1000 Men Raid Grocery Store and Seize Food," *Daily Ardmoreite*, January 20, 1931, and "Grocery Store Stormed and Windows Broken by Oklahoma City Mob," *Dallas Morning News*, January 21, 1931.

13. Los Angeles: "Use Tear Bombs on Radicals," *Dallas Morning News*, January 21, 1931. San Francisco: "Reds Start Food Riot," *Los Angeles Times*, February 2, 1931. El Paso: "Threat Made to Blow Up City Hall," *Fort Worth Star-Telegram*, September 30,

1931. Amarillo: "Amarillo Jobless Men Marched City Hall Wednesday," *Corsicana Daily Sun,* February 11, 1931.

14. Paris: "American National at Paris Closes Doors," *Fort Worth Star-Telegram,* February 6, 1931. Athens: "Athens Bank Closes for Liquidation," *Orange Leader,* March 11, 1931. Ringling: "They Say," *Daily Ardmoreite,* June 14, 1931. "Bank Closes Doors," *Fort Worth Star-Telegram,* July 4, 31. Enid: "Enid Bank Closes," *Miami Daily News,* July 27, 1931. Abilene: "Abilene Bank Closes, Plans Reorganization," *Fort Worth Star-Telegram,* August 30, 1931. Archer City: "Archer City Bank Closes, Banker Shot," *Breckinridge American,* September 25, 1931. Austin: "Austin Bank Closes, Few Calls for Money," *Dallas Morning News,* October 31, 1931.

15. On January 27, 2014, I met with fourteen lifetime residents of Red River communities to discuss life in Denison, Colbert, and surrounding areas during the Depression years, a meeting organized by the Colbert Historical Society. Their descriptions of life, conditions, and emotions on both sides of the river in the 1930s inform this chapter and later chapters. I also relied on interviews in the American Life Histories from the Folklore Project of the Federal Writers' Project, Library of Congress, Manuscript Division, WPA Federal Writers' Project Collection, and on newspapers from the period. There is a broad selection of books that describe family and commercial life in the 1930s, and some can be found in the sources section.

16. "Bridges at Four N. Texas Gateways Near Completion," *Dallas Morning News,* March 16, 1931.

17. "Puzzling Barriers," *Daily Oklahoman,* April 7, 1931.

18. "Texas-Oklahoma Bridge Pact Hit," *Daily Ardmoreite,* May 12, 1931.

19. "Red River Bridge Completed," *Denison Herald,* March 2, 1931.

20. "Puzzling Barriers," *Daily Oklahoman,* April 7, 1931.

21. "Use of Rangers at River Bridge Is Threatened," *Daily Ardmoreite,* April 2, 1931.

22. "Texas Highway Board Decided," *Daily Ardmoreite,* April 3, 1931.

23. "To Build Approaches to Red River Bridge," *Dallas Morning News,* April 13, 1931.

24. "To Open Red River Bridge About July 1," *Denison Herald,* May 18, 1931.

25. "Continue Work Opening Program New Free Bridge," *Durant Weekly News,* July 3, 1931.

26. This quote is from advertising appearing in the *Denison Herald,* June 16, 1931.

27. Byron C. Utect, Austin correspondent for the *Fort Worth Star-Telegram,* was probably the source. Utect had covered the capital for twenty years and was on a first-name basis with everyone from the capitol doorman to the Speaker of the House. Utect and the *Star-Telegram* led the press coverage of the bridge contracts. Unfortunately, there is no confirmation of Utect's role in the Byron C. Utect Papers, Special Collections, University of Texas at Arlington Libraries.

28. "Probe Asked of Bridge Deal," *Fort Worth Star-Telegram*, June 12, 1931.

29. Searches of the Texas State Library and Archives failed to find the original contract between the Texas Highway Commission and the Red River Bridge Company. However, clear photocopies of the signed contract are appended to most of the legal filings prepared by the Texas Attorney General's office. I used the contract included in *J. R. Handy vs. Johnson*, et al. (#348), June 24, 1931, Litigation Files, Records, Attorney General's Office, TSLAC.

30. Tilley had left the attorney general's office at the end of Governor Moody's administration. He was currently the legislative counsel to the Texas Senate.

31. "Probe Asked of Bridge Deal," *Fort Worth Star-Telegram*, June 12, 1931.

32. "No Effort to Conceal, Say Road Officials," *Fort Worth Star-Telegram*, June 14, 1931.

33. "Texas Highway Commission Faces Embarrassing Situation," *Durant Weekly News*, June 12, 1931.

34. *J. R. Handy vs. Johnson*, et al. (#348), telegram, J. W. Bailey to Highway Commission, June 24, 1931, Litigation Files, Records, Attorney General's Office, Texas State Library and Archives Commission (TSLAC).

35. *J. R. Handy vs. Johnson*, et al. (#348), letter, J. W. Bailey to J. V. Allred, June 25, 1931, Litigation Files, Records, Attorney General's Office, TSLAC.

36. "Red River Bridges," *Dallas Morning News*, June 30, 1931.

37. "May Appeal to Court," *Laredo Times*, June 30, 1931.

38. "Continue Denison Bridge Injunction," *Fort Worth Star-Telegram*, July 4, 1931.

39. "Texas Blocks Free Span," *Denison Herald*, July 3, 1931.

40. "Showdown on Span Muddle Due Saturday," *Denison Herald*, July 9, 1931, and "Time for Settlement," *Sherman Democrat*, July 3, 1931.

41. "Committee Is Making Plans for Span Fete," *Denison Herald*, July 6, 1931, and "Elaborate Fireworks Display to Feature Free Span Celebration," *Denison Herald*, July 7, 1931.

42. *J. R. Handy vs. Johnson*, et al. (#348), order, July 10, 1931, Litigation Files, Records, Attorney General's Office, TSLAC.

43. "Bridge Celebration Plans Are Cancelled," *Daily Ardmoreite*, July 12, 1931.

Chapter 7

1. "Cost of Bread Draws Attack from Murray," *Daily Oklahoman*, July 4, 1931.

2. "Senator Soaper Says—," *Los Angeles Times*, June 21, 1931.

3. "Farm to Market Road Plan Wins Approval," *Sallisaw Democrat-American*, July 9, 1931, and "Farm to Market Roads Designed," *Daily Oklahoman*, July 14, 1931.

4. "Oil Price Cuts Widen" and "Producers Strike in Two Oil Fields," both *New York Times*, July 10, 1931, and "Poll of House Shows Majority For Proration," *Fort Worth Star-Telegram*, June 12, 1931.

5. "Denison City Administration Practices What It Preaches," *Denison Herald*, April 5, 1931, and "Thousands Visit Cheese Factory," *Denison Herald*, April 25, 1931.

6. "To Erect Five-Story Factory Building," *Denison Herald*, April 27, 1931.

7. Ross S. Sterling and Edward W. Kilman, *Ross Sterling, Texan: A Memoir by the Founder of Humble Oil and Refining Company* (Austin: University of Texas Press, 2007), 126.

8. "Plan of Relief Is Adopted at Mass Meeting," *Dallas Morning News*, October 17, 1931.

9. Sterling and Kilman, *Ross Sterling, Texan*, 142.

10. Sterling and Kilman, *Ross Sterling, Texan*, 182.

11. "Red River Controversy," *Dallas Morning News*, July 13, 1931.

12. The tar baby is a folkloric character with roots in many cultures. Sterling would have been familiar with *Uncle Remus, His Songs and His Sayings* (Joel Chandler Harris, 1880). One of the stories tells of Br'er Rabbit's encounter with a doll made of tar and turpentine—a tar "baby." Thinking the doll is real, Br'er Rabbit strikes it, and his paw becomes stuck in the tar. The more the rabbit fights the tar baby, the more entangled and trapped he becomes.

13. "Delay Action on Span Case until July 11," *Denison Herald*, July 5, 1931.

14. Texarkana: "Nitro-Glycerine Tears Up Middle Span; Heavy Loss," *Hope Star* (Hope, AR), September 3, 1930. Gainesville: "Inquest to Be Held into Fatal Shooting," *Fort Worth Star-Telegram*, July 6, 1931. Lights and guards: "Understructure of Free Bridge Is Floodlighted at Night to Aid Guards," *Sherman Democrat*, July 11, 1931.

15. Much of the Easterly biographical information comes from Robert E. Hays, "Old School Journalist: A Biography of Claud Easterly," Commerce, TX: (master's thesis, East Texas State University, 1972), written from a series of recorded interviews with Easterly himself. "1922 Strike Helped Shape Denison" by R. C. Vaughn appeared in the *Sherman Herald-Democrat*, May 30, 2009, and also contained helpful biographical information.

16. "Negro, Locked in Vault, Cremated by Sherman Mob, Courthouse Burned, Troops Pushed by Lynchers," *Dallas Morning News*, May 10, 1930.

17. Chickasha: "Lynch Law Claims Life of Prisoner in Chickasha Jail," *Daily Ardmoreite*, June 1, 1930. Honey Grove: "Bullet-Filled Body of Negro Slayer Is Burned by Crowd," *Dallas Morning News*, May 17, 1930.

18. "Citizens Lament Delayed Opening of Free Bridge," *Denison Herald*, July 8, 1931.

19. Description of the bridge comes from "Plan Opening of Red River Bridge," *Dallas Morning News*, June 1, 1931.

20. "Citizens Lament Delayed Opening of Free Bridge."

21. The story is from Gordon Hines, *Alfalfa Bill: An Intimate Biography* (Oklahoma City: University of Oklahoma Press, 1932), 290. Hines quoted Murray,

telling this story in a dialect that is distracting and objectionable today. I've related the story as it was told, but without the dialect or racial references.

22. "Highways to Toll Bridges Blocked by Murray's Order," *Daily Oklahoman*, July 17, 1931.

23. "Governor Blames Texas for Span Delay," *Denison Herald*, July 17, 1931, and "Highways to Toll Bridges Blocked by Murray's Order," *Daily Oklahoman*, July 17, 1931.

24. One of the four original letters (perhaps the one given to Charley Davis), ended up in Ross Sterling's public papers. Scrawled across the bottom of the letter is "This is Gov. Murray's latest brain storm." From "Red River Bridge Controversy," July 17, 1931, Records of Governor Ross. S. Sterling, Texas State Library and Archives Commission (TSLAC).

25. "Motorists Have Inning as Free Span Is Temporarily Opened; Many Cross Bridge," *Denison Herald*, July 17, 1931.

26. "Murray's Actions Climax Bitter Span Controversy," *Denison Herald*, July 17, 1931.

27. "Troops Ready for Governor," *Fort Worth Star-Telegram*, July 17, 1931.

28. "Highway Department," July 17, 1931, Records of Governor Ross S. Sterling, TSLAC.

29. A concise history of the Texas State Militia is available from the Texas State Historical Association: Bruce A. Olson, "Texas National Guard," *Handbook of Texas Online* (http://www.tshaonline.org/handbook/online/articles/qnt02)

Chapter 8

1. "Hickman and Sterling on Job at Span," *Denison Herald*, July 17, 1931, and "Tom Hickman Heads Guards at Crossing," *Sherman Democrat*, July 17, 1931.

2. "State Rangers Stop Traffic over Free Project," *Austin American*, July 17, 1931.

3. This history of the Texas Rangers comes from two works by Mike Cox: *The Texas Rangers* (New York: Forge, 2008), and *Time of the Rangers: Texas Rangers from 1900 to the Present* (New York: Forge, 2009).

4. From his autobiography: William Warren Sterling, *Trails and Trials of a Texas Ranger*. (Norman, OK: University of Oklahoma Press, 1979). Also, "Captain Sterling True to Type of Rangers," *Dallas Morning News*, January 19, 1931.

5. "Hickman and Sterling on Job at Span."

6. Highway Department," telegram draft, July 18, 1931, Records of Governor Ross. S. Sterling, Texas State Library and Archives Commission (TSLAC).

7. "Annual Review Is Attended by Gov. Sterling," *Fort Worth Star-Telegram*, July 18, 1931.

8. Judge Newman was a Sequoyah Convention delegate and a long-time friend of Governor Murray's. The governor said he had been "tipped confidentially" that

the bridge company would seek a hearing in Carter's court. "Governor Murray Blames Texas for Span Delay," *Denison Herald*, July 17, 1931.

9. "Guardsmen to Be Used, Says Gov. Murray," *Denison Herald*, July 17, 1931.

10. "Executive Sends Crew to Tear All Barriers," *Sherman Democrat*, July 17, 1931.

11. "Gov. Murray Tells Statesman He'll Defend His Half of Span," *Austin Statesman*, July 17, 1931.

12. "Loy Differs with Sterling," *Denison Herald*, July 21, 1931.

13. "Rangers Close Free Bridge and Guard It," *Dallas Morning News*, July 18, 1931.

14. "Another Suit in Toll Case Is Developing," *Denison Herald*, July 17, 1931.

15. "Press Group Kept on Jump Investigating Many Rumors Regarding Bridge Squabble," *Denison Herald*, July 21, 1931.

16. Biographical information from Texas Ranger Hall of Fame and Museum, Waco, Texas. Likeable and unassuming as Hickman may have appeared, not everyone was a fan. One newspaper editor referred to Hickman as "a ranger with a string of dead men behind his unerring gat, and who has a reputation as a killer." "Comment and Censure by the Editor," *Durant Weekly*, July 24, 1931.

17. "Rangers Close Free Bridge and Guard It."

18. "Press Group Kept on Jump Investigating."

19. "Highway 75 Caravan Will Be Big Affair," *Denison Herald*, July 17, 1931.

20. "Employment Declines for June, Says Report," *Denison Herald*, July 17, 1931.

21. "The Bridge Climax," *Denison Herald*, July 17, 1931.

22. "Bridge War Leaves Traffic in Snarl; Court Help Asked," *Daily Oklahoman*, July 18, 1931.

23. "Highways to Toll Bridges Blocked on Murray's Order," *Daily Oklahoman*, July 17, 1931, and "Murray Upheld by Washington Officials," *Daily Oklahoman*, July 18, 1931.

24. "Bridge Opening Parley Fails to Get Results," *Austin American-Statesman*, July 21, 1931.

25. "Alfalfa Bill Leads Bridge Opening Fight," *Austin American-Statesman*, July 19, 1931. The Fort Worth and Houston editorials were reprinted for Oklahoma readers. See "What They Say about It in Texas," *Daily Oklahoman*, July 19, 1931.

26. The strong words between Loy and Sterling carried over to the next day and prompted questions by reporters. Loy later issued a statement. See the full statement at "Senator Loy Replies to Criticism from Sterling," *Fort Worth Star-Telegram*, July 21, 1931. Loy's hometown newspaper was a little more sensational, carrying the story under the subhead "Loy Is Flayed by Governor." See "Sterling and Allred Differ on Next Move," *Denison Herald*, July 21, 1931.

27. "Parley Called to Work Out Agreeable Plan," *Austin Statesman*, July 18, 1931.

28. The Sherman newspaper was a bit more restrained in its characterization of the dustup. See "Loy Differs with Sterling," *Sherman Democrat,* July 21, 1931.

29. "Jake Loy Says Governor Sterling Has Called Highway Commission Meeting," *Sherman Democrat,* July 18, 1931, and "Parley Called to Work Out Agreeable Plan," *Austin Statesman,* July 18, 1931.

30. "Murray Leads in Governors' Bridge Battle," *Dallas Morning News,* July 19, 1931.

31. Carter County: "Springer Man Made Head of Organization," *Daily Ardmoreite,* July 19, 1931. Lewis: "Victory Near for Murray in Toll War," *Daily Oklahoman,* July 19, 1931.

32. "Texans Want Bridges Opened, Murray Says," *Daily Ardmoreite,* July 19, 1931.

33. "State Relief Promised for Needy Miners," *Daily Oklahoman,* July 19, 1931, and "300 Jobless Men Demand Food; Get It," *Dallas Morning News,* July 19, 1931.

34. "State Relief Promised for Needy Miners," *Daily Oklahoman,* July 19, 1931.

35. "Alfalfa Bill Leads Bridge Opening Fight," *Austin American-Statesman,* July 19, 1931.

36. "Murray Leads in Governors' Bridge Battle," *Dallas Morning News,* July 19, 1931.

37. Ibid.

Chapter 9

1. "Hundreds of Persons See Span Sunday," *Denison Herald,* July 21, 1931.

2. "Press Group Kept on Jump Investigating Many Rumors Regarding Bridge Squabble," *Denison Herald,* July 21, 1931.

3. "Hundreds of Persons See Span Sunday," *Denison Herald,* July 21, 1931.

4. Several days later, an Associated Press reporter described a trick shooting demonstration by the rangers. The story had Texas Ranger Bob Goss splitting cards held edge-on with the pistol inverted; Tom Hickman was said to have hip-shot eighteen out of twenty matches at fifty feet. (See "Scheme to Open Red River Bridge Drawn at Austin," *San Antonio Express,* July 23, 1931.) Newspapers and magazines across the country picked up the story. The trick shooting demonstrations, however, likely never happened. Hickman admitted in a letter to the editor, "I am such a poor marksman that I never shoot until the other fellow has shot at me." The story was made up by reporters, he said. Goss could hit a playing card only "about twice out of every five shots." (See Letter to the Editor, *Time,* August 31, 1931.) A competitive shooter agreed such shooting was almost impossible. "There are 10,000 shooters in the U. S. who would pay good money to see that," he said. (See "Expert Pistol Shot Ready to Pay Cash to See Such Shooting," *Daily Oklahoman,* July 23, 1931.)

5. "Comment and Censure by the Editor," *Durant Weekly,* July 24, 1931.

6. "Hundreds of Persons See Span Sunday."

7. The account of this meeting comes from the reporting of veteran Austin correspondent Byron Utecht. "Efforts to Open Denison Bridge Fail," *Fort Worth Star-Telegram*, July 21, 1931.

8. In his narrative, Johnson says Oklahoma officials were party to negotiating and drafting the payment contracts, which were then sent to Oklahoma City for signatures. Lew Wentz, chairman of the Oklahoma Highway Commission, reneged on the agreement and Texas was forced to go it alone, Johnson says. This account varies considerably from contemporary newspaper accounts and existing records. Austin and Oklahoma City newspapers both report that the Oklahoma contingent left the negotiating session at the end of the first day; Texas and the bridge company negotiated for another day and a half. There is no existing record or correspondence in Texas or Oklahoma archives indicating that Oklahoma was an original party to the bridge agreements. Nowhere in Joe Bailey's lengthy legal filings or arguments does he mention that Oklahoma was ever a party to the bridge contracts. Six months later, Johnson would again describe events surrounding the contracts. The later statement, this time under oath, made no mention of Oklahoma's involvement with the contracts.

9. The quote is from a telegram sent to the newspaper by Senator Loy. "Sterling and Allred Differ on Next Move," *Denison Herald*, July 21, 1931.

10. It is unlikely Sterling could have authorized *any* amount on his own. The attorney general's opinion that the contracts were invalid would prevent the state treasurer from releasing any funds. Also, Loy's original bridge bill limited interstate bridge expenditures to half the cost of a new bridge and made no provision for acquisition or settlement costs.

11. The quote is from a telegram sent to the newspaper by Senator Loy. "Sterling and Allred Differ on Next Move," *Denison Herald*, July 21, 1931.

12. "Efforts to Open Denison Bridge Fail," *Fort Worth Star-Telegram*, July 21, 1931.

13. Sherman: "Call Mass Meet on Span," *Denison Herald*, July 20, 1931. Denison: "Call Mass Meeting Tonight," *Denison Herald*, July 21, 1931.

14. "Conference to Settle Bridge Row Resumed," *Dallas Times Herald*, July 20, 1931.

15. A description of the Sherman meeting (including most speeches) is at "Mass Meeting Demands Free Span Opening," *Denison Herald*, July 21, 1931.

16. A typescript of the Sherman resolution (including Gafford's attachment) is at "Red River Bridge Controversy," resolutions, July 21 and 22, 1931, Records of Governor Ross. S. Sterling, Texas State Library and Archives Commission (TSLAC).

17. A description of the Denison meeting (including most speeches) is at "Citizens, at Mass Meeting, Demand Immediate Opening of Barricaded 'Free' Span," *Denison Herald*, July 22, 1931.

18. "Murray Renews Bridge Fight," *Fort Worth Star-Telegram*, July 22, 1931.

19. A newspaper editor in a neighboring county applauded the idea of Grayson County seceding from Texas and joining Oklahoma—but only if Governor Murray were to pay the entire cost of the new bridge. "Worth a Smile," *Bonham Favorite*, July 23, 1931.

20. A typescript of the Denison resolution is at "Red River Bridge Controversy," resolutions, July 21 and 22, 1931, Records of Governor Ross. S. Sterling, TSLAC.

21. "The Issue of Tolls," *Daily Oklahoman*, July 18, 1931, and "Oklahoma Uses Ax," *Dallas Morning News*, July 18, 1931.

22. "But Will They Do It?" *Daily Ardmoreite*, July 21, 1931, and "Walking the Plank with the Editor," *McIntosh County Democrat*, July 23, 1931.

23. Martens: "They Say," *Daily Ardmoreite*, July 22, 1931. Davis: Letters from Readers, *Dallas Morning News*, July 21, 1931.

24. "'Alfalfa Bill' Calls Women to Bring Peace by Sewing Bees at Troop-Guarded Bridges," *New York Times*, July 22, 1931; "Women Asked to Join Battle over Bridges," *Binghamton Press* (Binghamton, NY), July 22, 1931; "Oklahoma Women May Force Bridge," *Washington Post*, July 22, 1931; "Murray War Tactics Scored," *Los Angeles Times*, July 22, 1931; and "Jurisdiction over Texas Bank Claimed by Spanish Treaty," *Dallas Morning News*, July 22, 1931.

25. "Murray Renews Bridge Fight," *Fort Worth Star-Telegram*, July 22, 1931.

26. "Red River Bridge Controversy," press statement, July 21, 1931, Records of Governor Ross. S. Sterling, TSLAC.

27. "Henryetta Aid Drive Started," *Daily Oklahoman*, July 22, 1931.

28. "Murray Risks Legal Attack, Delays Rents," *Daily Oklahoman*, July 22, 1931.

29. "Five Arrested as Hungry Men Ask for Food," *Dallas Morning News*, July 22, 1931.

30. "Alleged Agitators Freed by Court," *Daily Oklahoman*, July 24, 1931.

31. "Young Hickory," *New York Times*, August 2, 1931.

32. Both telegrams are available at "Red River Bridge Controversy," telegrams, July 20, 1931, Records of Governor Ross. S. Sterling, TSLAC.

33. "Appeal Made for War Dept. to Open Span," *Denison Herald*, July 20, 1931.

34. Chamber: "Huge Revenue from Drivers May Be Lost," *Daily Oklahoman*, July 19, 1931. Ad: "Notice to Oklahoma-Bound Tourists," *Dallas Morning News*, July 23, 1921.

35. "Dallas Losing Tourist Travel while Deadlock on Free Bridge Exists," *Dallas Times Herald*, July 22, 1931.

36. "Bridge Row Sends Cupid to Madill," *Daily Ardmoreite*, July 21, 1931, and "Madill Gets Texas Nuptial Business," *Daily Oklahoman*, July 22, 1931. For the other side of the coin, see "Bridge Battle Hurts Marriage License Sales," *Durant Weekly*, July 24, 1931.

37. "The Case of Ray Terrill," *Dallas Morning News*, July 24, 1931.

38. "Texas Highway Department," Womack letter, July 23, 1931, Records of Governor Ross. S. Sterling, TSLAC.

39. "Red River Bridge Controversy," Womack letter, July 22, 1931, Records of Governor Ross. S. Sterling, TSLAC.

40. "Red River Bridge Controversy," Mills letter, July 22, 1931, Records of Governor Ross. S. Sterling, TSLAC.

41. "Adjutant General Dept.," Jones letter and clipping, July 18, 1931, Records of Governor Ross. S. Sterling, TSLAC.

42. "But Will They Do It?" *Daily Ardmoreite*, July 21, 1931.

43. Letter from Z. W. Davis of Colorado, Texas. Letters from Readers, *Dallas Morning News*, July 21, 1931.

44. "Red River Bridge Controversy," Tatum letter, July 23, 1931, Records of Governor Ross. S. Sterling, TSLAC.

45. "Plan to Speed Span Opening Is Agreed Upon," *Dallas Morning News*, July 23, 1931.

46. "Sherman Man Is Drowned in Red River," *Sherman Democrat*, July 21, 1931, and "Body of Carter Taken from Red River Thursday," *Denison Herald*, July 23, 1931.

Chapter 10

1. Byron Utecht, Austin correspondent for the *Fort Worth Star-Telegram*, provided some of the best news coverage of the day's events. "Senate Will Get Bill for Bridge Suit," *Fort Worth Star-Telegram*, July 23, 1931.

2. By Wednesday afternoon, Allred, Loy, and Sterling had given up hope for relief from the US Supreme Court. "Citizens, at Mass Meeting, Demand Immediate Opening of Barricaded 'Free' Span," *Denison Herald*, July 22, 1931.

3. Loy found time during the day to update editors of the newspapers in his district via telegram. The Bonham editor printed one of the wires in which Loy indicates he has lined up enough votes for the measure to pass. "Senate Bill No. 9 by Loy May Be the Way Out of the Matter," *Bonham Favorite*, July 23, 1931.

4. "Plan to Speed Span Opening Is Agreed Upon," *Dallas Morning News*, July 23, 1931.

5. "Toll Bridge Owners May Sue Oklahoma in Federal Court," *Sherman Democrat*, July 23, 1931.

6. "Gore Wisecracks to Legislators," *Dallas Morning News*, July 24, 1931.

7. Ibid.

8. *Journal of the Senate of Texas*, during the First Called Session of the Forty-Second Legislature, 65–66, 1931.

9. "Bills to Reopen Free Bridge Passed," *Austin Statesman*, July 23, 1931.

10. "Gore Aims Ridicule at Republicans," *Fort Worth Star-Telegram*, July 24, 1931.

11. "Denison Bridge Settlement Appears to Be Nearing End," *Bonham Favorite*, July 23, 1931.

12. "Van Zandt Says House Members Helped with Bill," *Sherman Democrat*, July 24, 1931.

13. "Measure Goes to Governor for Signature; When Free Bridge to Open Not Known," *Denison Herald*, July 24, 1931.

14. *Journal of the House of Representatives*, First Called Session of the Forty-Second Legislature, 59–61, 1931.

15. "Van Zandt Says House Members Helped with Bill," *Sherman Democrat*, July 24, 1931.

16. The order is at Declaration of Martial Law, July 22, 1931, box 1, folder 1, Record Group 8-I-4 Miscellaneous, Governor's Office Records, Oklahoma State Archives, Oklahoma Department of Libraries, Oklahoma City.

17. The full text of the order also appears at "Martial Law Order," *Daily Oklahoman*, July 24, 1931.

18. "Oklahoma Guardsmen Will Block Bridge as Governor Puts Strip under Martial Law," *Daily Oklahoman*, July 24, 1931; "Governor Murray Sends Oklahoma Troops to Guard Toll Bridge," *Sherman Democrat*, July 24, 1931; and "290 Men Ready for Martial Law Order," *Fort Worth Star-Telegram*, July 24, 1931.

19. "Texas Rangers Not Informed of Order," *Fort Worth Star-Telegram*, July 24, 1931.

20. "Adj. Gen. Sterling Returns to Austin," *Fort Worth Star-Telegram*, July 24, 1931.

21. "Chamber of Commerce Tries to Get Red River Bridge Open in Time for Highway 75 Motorcade Saturday," *Sherman Democrat*, July 25, 1931.

22. "Lights Placed on Rail, Toll Spans in Safety Effort," *Denison Herald*, July 24, 1931.

23. "Murray Sends Guardsmen to Bottle Bridge," *Denison Herald*, July 24, 1931.

24. "290 Men Ready for Martial Law Order," *Fort Worth Star-Telegram*, July 24, 1931.

25. "Oklahoman Is Driven Away," *Sherman Democrat*, July 24, 1931, and "Sidelights at the Battlefront," *Denison Herald*, July 24, 1931.

Chapter 11

1. "Toll Bridge at Durant Closed," *Daily Ardmoreite*, July 24, 1931.

2. "Gov. Murray Sends Oklahoma Troops to Guard Toll Bridge," *Sherman Democrat*, July 24, 1931.

3. Lawrence Foster was one of the Oklahoma National Guardsmen at the toll bridge in July 1931. He was interviewed before his death in 2007. "Texas Rangers, Oklahoma Guard, Meet in Red River Showdown," *Daily Oklahoman*, April 18, 1999.

4. "Toll Bridge at Durant Closed," *Daily Ardmoreite*, July 24, 1931.

5. "Armed Forces of Two States Guarding Red River Bridges," *Dallas Morning News*, July 25, 1931.

6. The eight-page wire is at "Red River Bridge Controversy," J. W. Bailey telegram, July 24, 1931, Records of Governor Ross. S. Sterling, Texas State Library and Archives Commission (TSLAC).

7. The first called special session of the forty-second Texas legislature was due to end at midnight, July 31. If the waiver legislation had not passed the full legislature by then, Loy's next opportunity would be two months away, at best.

8. The bill's passage through both houses of the legislature that morning is recorded at *Journal of the House of Representatives*, First Called Session of the Forty-Second Legislature, 68–69, 72–73, and *Journal of the Senate of Texas* during the First Called Session of the Forty-Second Legislature, 112–113.

9. For the full text of the final bill, see Hans Peter Mareus Neilsen Gammel, *Laws of Texas 1822–1897* (Austin: Gammel Book Publishing, 1898), 28:3–5.

10. "Plane Will Fly Proposal to Sterling in Houston," *Austin Statesman*, July 24, 1931.

11. "Sidelights at Battlefront," *Denison Herald*, July 26, 1931, and "Armed Forces of Two States Guarding Red River Bridges," *Dallas Morning News*, July 25, 1931.

12. "Armed Forces of Two States Guarding Red River Bridges," *Dallas Morning News*, July 25, 1931.

13. "Judge Neblett to Preside," *Dallas Morning News*, July 24, 1931.

14. "Armed Forces of Two States Guarding Red River Bridges," *Dallas Morning News*, July 25, 1931.

15. Red flags: "Wanted, an Enemy or So to Keep Army of 36 Warriors Occupied," *Daily Oklahoman*, July 25, 1931. Mail: "Mails Exempt," *Dallas Morning News*, July 24, 1931. Soldiers: "Texas Rangers, Oklahoma Guard, Meet in Red River Showdown," *Daily Oklahoman*, April 18, 1999.

16. "Armed Forces of Two States Guarding Red River Bridges," *Dallas Morning News*, July 25, 1931.

17. "Texas Rangers, Oklahoma Guard, Meet in Red River Showdown," *Daily Oklahoman*, April 18, 1999, and "Sidelights at Battlefront," *Denison Herald*, July 26, 1931. Also, my interview with James Hannan, October 22, 2014.

18. "Bridge Bill Is Passed, Signed," *Fort Worth Star-Telegram*, July 25, 1931.

19. Ibid.

20. "Oklahoma Corps Wins Bridgehead in Broad Daylight," *San Antonio Express*, July 25, 1931.

21. "Road Sponsors Seek Removal of Injunction," *Dallas Times Herald*, July 25, 1931.

22. "Chamber of Commerce Tries to Get Red River Bridge Open in Time for Highway 75 Motorcade Saturday," *Sherman Democrat*, July 25, 1931.

23. "Guards at Bridge Add Phonograph," *Austin American-Statesman*, July 25, 1931. Music choices: "Sidelights at Battlefront," *Denison Herald*, July 26, 1931.

24. "Toll Bridge at Durant Closed," *Daily Ardmoreite*, July 24, 1931.

Chapter 12

1. "Injunction Suspended," *Sherman Democrat*, July 26, 1931.

2. "Injunction Is Suspended until Aug. 3 Hearing," *Fort Worth Star-Telegram*, July 26, 1931.

3. "Chamber of Commerce Tries to Get Red River Bridge Open in Time for Highway 75 Motorcade Saturday," *Sherman Democrat*, July 25, 1931.

4. Ibid., and "Highway 75 Motorcade among First over Red River Free Bridge Saturday," *Sherman Democrat*, July 26, 1931.

5. "Loy Declares This Is Happy Day for Texas," *Denison Herald*, July 26, 1931.

6. "Murray Is Defiant in Face of US Injunction," *Fort Worth Star-Telegram*, July 25, 1931.

7. "Guardsmen Receive Some Courtesies," *Durant Weekly News*, July 31, 1931.

8. "Troops While Away Hours at Toll Bridge," *Fort Worth Star-Telegram*, July 25, 1931.

9. "Herald Reporter and Fiancée Are Honored at Fete," *Denison Herald*, July 21, 1931, and "Easterly-Davis Wedding Service," *Denison Herald*, July 27, 1931. See also Robert E. Hays, "Old School Journalist: A Biography of Claud Easterly" Commerce, TX: (master's thesis, East Texas State University, 1972).

10. Bailey's appearance and demeanor in court that morning are from observations by the United Press and *Daily Oklahoman* reporters. "Murray Assumes Actual Command of River Troops," *Daily Oklahoman*, July 26, 1931, and "Oklahoma Bridge Closing Enjoined," *San Antonio Express*, July 26, 1931.

11. Bailey's words, as quoted throughout this section, are taken from the record of his lawsuit. See J.R. Handy et al. v. Burt McCaleb et al. [4340–USDist. Ct.–OK], petition, July 25, 1931, Litigation Files, Records, Attorney General's Office, Texas State Library and Archives Commission (TSLAC); and "Oklahoma Bridge Closing Enjoined," *San Antonio Express*, July 26, 1931.

12. "Hundreds Watch as Rangers Order Bridge Barriers Torn Away," *Sherman Democrat*, July 26, 1931.

13. "Blocked Bridges Are Open to Planes Which Take Passengers Over," *Denison Herald*, July 24, 1931, and "Armed Forces of Two States Guarding Red River Bridges," *Dallas Morning News*, July 25, 1931.

14. See "Injunction Suspended," *San Antonio Express*, July 26, 1931, for the complete 102-word text of Judge Kennerly's suspension order.

15. "Injunction Suspended," *Sherman Democrat*, July 26, 1931.

16. In news accounts of the day (and later), Governor Sterling, Gibb Gilchrist, and Sidney Benbow all claimed credit for having "given the order" to open the

Red River bridge to traffic. As described here, all of them were somewhat correct. However, Tom Hickman said at the time that he acted only when he received a direct order from Adjutant General W. W. Sterling.

17. "Loy Declares This Is Happy Day for Texas," *Denison Herald*, July 26, 1931.

18. "Hundreds Watch as Rangers Order Bridge Barriers Torn Away," *Sherman Democrat*, July 26, 1931.

19. "Loy Declares This Is Happy Day for Texas."

20. "Hundreds Watch as Rangers Order Bridge Barriers Torn Away."

21. "Loy Declares This Is Happy Day for Texas."

22. "Governor Orders Rangers Withdrawn," *Sherman Democrat*, July 26, 1931. For photos of Tom Hickman posing with some of the earliest bridge crossers and of the Texas Rangers breaking camp, see "Barriers Removed on Free Bridge," *Dallas Morning News*, July 26, 1931.

23. The quotes by Judge Neblett from his order are at "Murray Assumes Actual Command of River Troops," *Daily Oklahoman*, July 26, 1931, and "Murray, with Horse Pistol, Commands Troops at Bridge," *Fort Worth Star-Telegram*, July 26, 1931.

24. "Murray Assumes Actual Command of River Troops."

25. "'US Can't Cite Me, Let 'Em Try,'—Murray," *Daily Oklahoman*, July 26, 1931.

26. "Murray Assumes Actual Command of River Troops."

27. "Will Go in Person to Toll Bridge to Direct Keeping Up Blockade," *Sherman Democrat*, July 26, 1931.

28. "'War' but No Foe," *Kansas City Star*, July 26, 1931.

29. "The First Field General," *Daily Oklahoman*, July 26, 1931.

30. "Murray Assumes Actual Command of River Troops." For a photo of a wobbly W. C. Lewis, just off the airplane and still wearing his leather flying helmet, serving Murray with Judge Neblett's order, see "Oh, for the Life of an Army Lad at Murray Barracks on the River Front," *Daily Oklahoman*, July 27, 1936.

31. "'US Can't Cite Me, Let 'Em Try,'—Murray."

32. The *Daily Oklahoman* published the most complete and colorful coverage of Murray's dash to the Red River toll bridge. If there wasn't a reporter in the car with Murray, there was certainly one following directly behind the governor every mile of the way.

33. Reporters from the *Daily Oklahoman* and Associated Press and a stringer for the *New York Times* provided first-person coverage of Bailey's encounter with Colonel MacDonald. See "Murray Assumes Actual Command of River Troops," *Daily Oklahoman*, July 26, 1931; "An Officer Defies Court," *Kansas City Star*, July 26, 1931; and "Alfalfa Bill Arms, Goes to Bridge War," *New York Times*, July 26, 1931.

34. "Murray Amuses Enjoining Judge," *San Antonio Express*, July 26, 1931, and "Murray Hurls Defiance at a Federal Judge," *Kansas City Star*, July 26, 1931.

35. "'Cub' Reporter Recalls Red River Bridge War," *Sherman Democrat*, November 17, 1985.

36. "'US Can't Cite Me, Let 'Em Try,'—Murray."

37. "Probe of US Span Orders Sought," *Daily Oklahoman*, July 27, 1931.

38. One observer said Murray "was greeted with lusty yells" by the guardsmen. See "Alfalfa Bill Arms, Goes to Bridge War."

39. "Murray Assumes Actual Command of River Troops."

40. "Governor Murray Takes Command of Bridge Troops; Will Defy US Injunction," *Dallas Times Herald*, July 26, 1931.

41. "Easterly-Davis Wedding Service," *Denison Herald*, July 27, 1931.

42. "Sidelights at Battlefront," *Denison Herald*, July 26, 1931.

43. This informal, late-night, one-more-question session is described at length by a *Daily Oklahoman* staff reporter. See "'US Can't Cite Me, Let 'Em Try,'—Murray."

44. In Gordon Hines, *Alfalfa Bill: An Intimate Biography* (Oklahoma City: University of Oklahoma Press, 1932), the governor's campaign biography, the author says Murray had "observed a gradual usurpation of unconstitutional authority by minor federal courts" (288). His purpose in opening the free bridge, the author implies, was to push back at the courts. Alfalfa Bill says as much himself in William H. Murray, *Memoirs of Governor Murray and True History of Oklahoma: Together with His Biography, Philosophy, Statesmanship, and Oklahoma History, Interwoven* (Boston: Meador, 1945), when he discusses only the rightness of his constitutional reasoning, and not the events of the bridge war itself. A reading of the daily record of events occurring at the time, however, indicates that Murray touched off the Red River Bridge War as a populist stratagem, and only later hung his coat on the constitutional peg (456–501).

Chapter 13

1. Members of the press "were about as thick as mosquitoes along the battle front" that weekend, according to Easterly. Among others, he mentions "visitors from New York" and Chicago. See "Sidelights at Battlefront," *Denison Herald*, July 26, 1931. The "Alfalfa Bill Barracks" dateline appears on "Posing Takes All Murray's Battle Time," *Daily Oklahoman*, July 27, 1931.

2. "'US Can't Cite Me, Let 'Em Try,'—Murray," *Daily Oklahoman*, July 26, 1931.

3. "Sterling Pleased Bridge Spat Ends," *San Antonio Express*, July 26, 1931.

4. "Posing Takes All Murray's Battle Time." The newspaper rendered the "cross another creek" story in a racial dialect that would be objectionable today. I've kept the quote, but changed words and spellings to American English.

5. "Probe of US Span Orders Sought," *Daily Oklahoman*, July 27, 1931.

6. "Toll Bridge Open; Troops Stay on Job," *Denison Herald*, July 27, 1931.

7. "Sidelights at Battlefront," *Denison Herald*, July 26, 1931.

8. "Probe of US Span Orders Sought."

9. "Posing Takes All Murray's Battle Time."

10. From my October 22, 2014, interview with James Hannan, who accompanied his father to Murray's headquarters on the river.

11. "Handshakers Now Claim Governor Did Them Honor," *Denison Herald*, July 29, 1931.

12. "Posing Takes All Murray's Battle Time."

13. "Probe of US Span Orders Sought."

14. "Posing Takes All Murray's Battle Time."

15. "Probe of US Span Orders Sought."

16. "Free Bridge Placed under Martial Law," *Dallas Times Herald*, July 27, 1931, and "Land Dispute Back of Late Murray Order," *Denison Herald*, July 27, 1931.

17. "Murray Puts Entire Span under Edict," *Denison Herald*, July 27, 1931.

18. "Alfalfa Bill Bows to Federal Court," *New York Times*, July 27, 1931.

19. "Traffic Permitted over Toll Structure," *McKinney Courier-Gazette*, July 27, 1931.

20. "Effort to Close Free Bridge Again Is Seen," *Fort Worth Star-Telegram*, July 28, 1931.

21. Declaration of Martial Law and Executive Order, July 25, 1931, box 1, folder 1, Record Group 8-I-4 Miscellaneous, Governor's Office Records, Oklahoma State Archives, Oklahoma Department of Libraries, Oklahoma City.

22. "Toll Bridge Open; Troops Stay on Job," *Denison Herald*, July 28, 1931.

23. "Troops Posted on Free Bridge," *Fort Worth Star-Telegram*, July 28, 1931.

24. "Plan Gigantic Barbecue in Honor of Gov. W. H. Murray," *Durant Weekly*, July 31, 1931.

25. Retailers: "Free Bridge Helping Denison in Big Way, Merchants Agree; Removes Old Trade Barriers," *Denison Herald*, August 9, 1931. Market Day: "Kennerly Decision Awaited Anxiously by People Hoping Bridge Will Not Be Closed," *Denison Herald*, August 2, 1931.

26. "Free Bridge to Boost Attendance at Southeastern," *Durant Weekly*, August 21, 1931, and "288 Marriage Licenses Issued during August," *Durant Weekly*, September 4, 1931.

27. Purcell toll bridge: "Detour Is Made across Stream under Pay Span," *Dallas Morning News*, July 29, 1931. Ice: "State Sale of Ice May Grow," *Daily Oklahoman*, July 28, 1931.

28. "Lull before Storm Seen in Bridge Battle; Await Court Injunction Ruling," *Denison Herald*, July 29, 1931.

29. "Oklahoma, Texas Near Civil War," *Bakersfield Californian*, July 18, 1931, and "Battle of the Red River," *Syracuse Herald*, July 19, 1931.

30. "Station H-E-R-A-L-D Broadcasting," *Denison Herald*, July 29, 1931. Also, see "New York Paper Approves Stand of Gov. Bill Murray," *Denison Herald*, August 7, 1931. The *New York Times* and the *Washington Post* carried stories about the bridge war and Governor Murray each day, from July 26, 1931, through July 30, 1931.

31. Newcastle and Oakland: "Red River Bridge Campaign Taken as Summer Diversion," *Newcastle News*, August 1, 1931, and "Neighbors Scrap," *Oakland Tribune*, July 21, 1931. Oshkosh: "Face to Face," *Oshkosh Daily Northwestern*, August 1, 1931.

32. The Murray photo appeared in the *Los Angeles Times* and the *Atlanta Journal-Constitution*, both on August 1, 1931. "Ready to Fight," *Rochester Times-Union*, August 2, 1931; "Bill Murray Is Oklahoma's Best," *Bluefield Daily Telegraph* (Bluefield, WV), August 9, 1931; "Horatius Still Guarding the Bridge," *Kansas City Star*, July 26, 1931.

33. "His Turn to Be Enjoined," *Dallas Times-Herald*, July 28, 1931.

34. "Governor Murray Comes," *Dallas Morning News*, August 15, 1931.

35. "The Situation Has Not Been Helped," *Dallas Times Herald*, July 19, 1931.

36. "A Dirt Conspiracy?" *Dallas Morning News*, July 25, 1931; "Some Texas Blunders," *Daily Ardmoreite*, July 26, 1931, and "Comment and Censure," *Durant Weekly*, July 24, 1931.

37. "Red River Bridge Controversy," *The United States Daily* letter, July 23, 1931, and *The United States Daily* letter, August 5, 1931.

38. Records of Governor Ross. S. Sterling, Texas State Library and Archives Commission (TSLAC).

Chapter 14

1. "Van Alstyne Merchant Shows Red River 'War' in Window," *Dallas Morning News*, August 2, 1931.

2. "Court Orders over Bridges Are Dissolved," *Dallas Morning News*, August 7, 1931, and "Free Bridge Helping Denison in a Big Way, Merchants Agree; Removes Old Trade Barriers," *Denison Herald*, August 9, 1931.

3. "Bridge Battle Is on Again as Company Sues," *Dallas Morning News*, October 21, 1931.

4. "Governor Murray Stirs the Mixture," *Fort Worth Star-Telegram*, July 20, 1931.

5. "The Situation Has Not Been Helped," *Dallas Times Herald*, July 19, 1931.

6. "Kennerly Decision Awaited Anxiously by People Hoping Bridge Will Not Be Closed," *Denison Herald*, August 2, 1931.

7. "Red River Bridge Case," *Dallas Morning News*, October 29, 1932, and "Texas River Bridge Hearing Set," *Daily Oklahoman*, November 4, 1931.

8. "Red River Bridge Company Seeking Heavy Damages as Result Opening Free Way," *Corsicana Sun*, January 11, 1932.

9. "Chairman Highway Commission Denies Alleged Contract," *Corsicana Sun*, January 12, 1932.

10. "Sterling Tells of Bridge Contract," *Fort Worth Star-Telegram*, January 13, 1932.

11. "Right of Texas Highway Commission to Enter into Bridge Contract Argued," *Corsicana Sun*, January 14, 1942.

12. "Bridge Company Loses Suit," *Corsicana Sun*, January 15, 1932.

13. According to later stories, Allred's assistant, Sidney Benbow, learned of the contracts weeks earlier while preparing a witness for trial. "State Highway Commission Paid Terral Company," *Corsicana Semi-Weekly Light*, January 19, 1932.

14. "Allred Flays Highway Body Settlement," *Fort Worth Star-Telegram*, January 16, 1932, and "Inquiry Court on Settlement Is Held by Allred," *San Antonio Light*, January 16, 1932.

15. "Red River Bridge Controversy," Stark letter, January 9, 1932, Records of Governor Ross. S. Sterling, TSLAC.

16. "Red River Bridge Controversy," Womack letter, January 11, 1932, Records of Governor Ross. S. Sterling, TSLAC.

17. "Governor of Texas Losing Supporters," *New York Times*, September 20, 1931.

18. "Appeal in Bridge Suit Is Dismissed," *Dallas Morning News*, February 11, 1932, and "Bridge War to Flare Again If Company Sues Oklahoma," *Daily Oklahoman*, March 17, 1932.

19. "Oklahoma Again Places Soldiers to Guard Bridge," *Dallas Morning News*, June 7, 1932.

20. This deployment was more modest than the earlier ones. Murray ordered only four men to the bridge and withdrew them after two days. "State to File New Briefs in Federal Case," *Daily Oklahoman*, and "Acts as State Loses Plea to Dismiss Suit," *Fort Worth Star-Telegram*, June 7, 1932.

21. William H. Murray Collection, Subject & Personality Materials, Rosser to Murray, April 11, 1933, Carl Albert Center Congressional Archives, University of Oklahoma.

22. "State Beaten in First Toll Bridge Battle," *Daily Oklahoman*, October 27, 1932.

23. "Red River Bridge Parley Arranged," *Fort Worth Star-Telegram*, November 18, 1932.

24. William H. Murray Collection, Subject & Personality Materials, Murray to Attorney General King, March 13, 1933, Carl Albert Center Congressional Archives, University of Oklahoma.

25. William H. Murray Collection, Subject & Personality Materials, Rosser to Murray, April 11, 1933, Carl Albert Center Congressional Archives, University of Oklahoma.

26. William H. Murray Collection, Subject & Personality Materials, Rosser to Murray, April 11, 1933, Carl Albert Center Congressional Archives, University of Oklahoma. The following quotes by Rosser are from this letter.

27. "Red River Bridge Company Is Given $168,000 Damages," *Dallas Morning News*, April 12, 1933.

28. William H. Murray Collection, Subject & Personality Materials, Rosser to Murray, April 13, 1933, Carl Albert Center Congressional Archives, University of Oklahoma.

29. "Judge Demeaned in Toll Bridge Case," *Fort Worth Star-Telegram,* July 15, 1933.

30. "State Beaten in Bridge War," *Daily Oklahoman,* July 4, 1933.

31. "Murray Defies US Judgement," *Daily Ardmoreite,* July 5, 1933.

32. "Murray's Span Fight Upheld," *Fort Worth Star-Telegram,* June 19, 1934. The court of appeals reversed Pollock on the narrow issue of rights of Chickasaw nationhood at the time of Oklahoma statehood. See State of Oklahoma v. Handy, 71 F.2d 697 (10th Cir.1934); decided August 27, 1934; #978.

33. "Red River Bridge Costs State $1,211," *Daily Ardmoreite,* January 31, 1936.

34. "Toll Bridge Purchase Asked," *Dallas Morning News,* May 10, 1936, and "Battle of Red River Bridge Comes to an End," *Ada Evening News,* November 2, 1937.

35. TxDOT, *Report on Texas Bridges as of September 2014.* Texas maintains the largest inventory of bridges in the nation—53,000 bridges for public vehicular traffic.

Epilogue

1. William H. Murray, *Memoirs of Governor Murray and True History of Oklahoma: Together with His Biography, Philosophy, Statesmanship, and Oklahoma History, Interwoven* (Boston: Meador, 1945), 2:499.

2. Ben R. Guttery, *Representing Texas: A Comprehensive History of US and Confederate Senators and Representatives from Texas* (Charleston, SC: BookSurge, 2008), 71, and "Death Takes Capt. Bailey, Attorney," *Dallas Morning News,* July 18, 1943.

3. "Death Takes Judge Loy of Sherman," *Dallas Morning News,* July 5, 1945, and "History of Loy Lake Park," *Herald-Democrat,* October 5, 1997.

4. "Claude Easterly, 91, Former Editor of Texas Paper," *Atlanta Journal-Constitution,* March 19, 1999.

5. "Baptists to Honor Kennerly," *Dallas Morning News,* July 30, 1962; "Federal Judge Colin Neblett Dies of Stroke," *Hobbs Daily News Sun* (Hobbs, NM), May 8, 1950; and John C. Pollock's Funeral Arranged," *Fort Worth Star-Telegram,* January 26, 1937.

6. Biographical entries for Sterling and Hickman are in the Texas Ranger Hall of Fame and Museum, Waco, TX. See also "Captain Sterling True to Type of Texas Rangers," *Dallas Morning News,* January 19, 1931.

7. "Final Tribute Paid John A. MacDonald," *Daily Ardmoreite,* November 12, 1939.

8. "Rice Tilley Sr., 90, Fort Worth Attorney, Dies," *Fort Worth Star-Telegram,* April 14, 1990.

9. "Ex-Federal Commissioner Sidney Benbow Dies at 71," *Victoria Advocate,* September 29, 1965.

10. "Oklahoma Attorney Named," *Dallas Morning News,* January 2, 1934, and "Three Lawyers in Family Make Court Bow," *Fort Worth Star-Telegram,* March 3, 1939.

11. Allred gubernatorial campaign ad: "Allred's Remarkable Record Withstands Bitter Attacks," *Rusk Cherokeean*, August 24, 1934.

12. "Governor of Texas Losing Supporters," *New York Times*, September 20, 1931.

13. Ross S. Sterling and Edward W. Kilman, *Ross Sterling, Texan: A Memoir by the Founder of Humble Oil and Refining Company* (Austin: University of Texas Press, 2007), 217.

14. Sterling and Kilman, *Ross Sterling, Texan*, 226–237.

15. Betty T. Chapman, "Diverse Legacy of Ross Sterling Still Evident All Over Houston," *Houston Business Journal*, September 14, 2001.

16. Settlers: "Murray Steps for 'Folks' at Weatherford," *Daily Oklahoman*, August 16, 1931. Dallas: "Preaches Doctrine of Common People to Crowd of 15,000," *Dallas Morning News*, August 16, 1931. Editorial: "He Shows His Wares," *Dallas Morning News*, August 18, 1931.

17. "'Alfalfa Bill' in War, Peace, and Politics," *The Literary Digest*, September 19, 1931; "Bayonet Business," *Collier's*, September 26, 1931; "Victuals and Battles," *Saturday Evening Post*, December 19, 1931; "Bread, Butter, Bacon, Beans," *Time*, February 29, 1932; "The Sage of Tishomingo," *American Mercury*, May 1932; "Presidential Possibilities," *The Nation*, April 6, 1932; and "Trend of the Week," *Outlook & Independent*, August 5, 1931.

18. "Bread, Butter, Bacon, Beans."

19. "Gov. Murray Formally Announces Candidacy," *Fort Worth Star-Telegram*, February 22, 1932.

20. "Gov. Murray Formally Announces Candidacy," *Fort Worth Star-Telegram*, February 2, 1932.

21. Tony Hillerman, *Seldom Disappointed: A Memoir* (New York: HarperCollins, 2001).

22. "Red River Bridge Case," *Dallas News*, October 29, 1932; and "Paragraphs by R. M.," *Daily Ardmoreite*, September 2, 1934.

23. Some could take exception to the "end of the private toll bridge era" statement, pointing to San Francisco's Golden Gate Bridge, then under construction. Construction was financed by a private group intending to raise money by selling bonds. For two years, as the Depression deepened, sales of the bonds languished. The project might have collapsed altogether if A. P. Giannini, Bank of America founder, hadn't stepped forward to subscribe to the whole lot. Because of financial issues, the state of California found it necessary to take over the bridge. See Kevin Starr, *Golden Gate: The Life and Times of America's Greatest Bridge* (New York: Bloomsbury, 2010).

24. James B. Reed, "On the Road (and Bridge) Again," *State Legislatures Magazine*, January 2015.

25. *Toll Facilities in the United States*, Publication No: FHWA-PL-13–037 (Washington, DC: Office of Highway Policy Information, Federal Highway Administration, September 2013).

26. Phineas Baxandall, *Private Roads, Public Costs—The Facts about Toll Road Privatization and How to Protect the Public* (Boston: US PIRG, 2009).

27. Traffic count comes from Texas Department of Transportation, *2012 Texas Traffic Flowband Map.*

28. Painting: "Texas-Oklahoma Reach Unity on Red River Bridge," *Dallas Morning News,* January 20, 1951. Companion bridge: "Texas, Oklahoma Feuded along Red River in 1931," *Dallas Morning News,* March 22, 1953.

29. "Span That Almost Touched Off War Is Near Collapse," *Daily Oklahoman,* August 31, 1958, and "Bridge over Red Closed For Always," *Daily Oklahoman,* March 13, 1960.

30. "It Took a Fire to Do the Job," *Dallas Morning News,* August 12, 1961.

31. Wyota Hannan spent eight years talking to officials and raising money to save the 1931 bridge, citing its value as a tourist attraction and it historical significance. See "History Buff Trying to Save Old Bridge," *Daily Oklahoman,* July 25, 1987. Demolition: "Crews Demolish Bridge on the Red River," *Dallas Morning News,* December 8, 1995, and "'Red River War' Bridge Surrenders to March of Time," *Daily Oklahoman,* December 10, 1995.

SOURCES

A Note on Sources

The full story of the Red River Bridge War extends far beyond a short stretch of the Red River and two weeks in July 1931. The conflict was decades in the making; it involved the governors, legislators, and citizens of two states. The war, its prelude, and its aftermath required tank cars of newspaper printers' ink to report and explain; it consumed months in courtrooms—state and federal—across four states.

The sources necessary to tell the story are just as far flung. The complete narrative of the Red River Bridge War resides in state and private archives and collections, in the recollections of those who contested the bridges, and in the contemporary written records of those who witnessed it. But those records alone provide insufficient testimony. Oklahoma National Guardsmen and Texas Rangers took up stations at the Red River bridges more than eighty years ago, a time quite distant from our own. To understand why two states armed up and went to war over a two-bit toll bridge in 1931, it is necessary also to appreciate the economic, cultural, and social context of the time.

Listed here are many of the sources I consulted in order to tell this story. Most of the source records were created contemporaneously with the events they describe; others were created later, after study and interpretation. I have occasionally commented on some of the sources, describing their special benefits or regrettable drawbacks.

Archives and Collections

Governor William H. Murray, Governor's Office Records, Oklahoma State Archives, Oklahoma Department of Libraries, Oklahoma City.

Executive Military Order, Martial Law, July 23, 1931.

Executive Military Order, Martial Law, July 27, 1931.

Ross Shaw Sterling Papers, 1930–1939, Dolph Briscoe Center for American History, University of Texas at Austin.

Scrapbook, containing miscellaneous clippings from *Denison Herald*, *Sherman Democrat*, *Bonham Favorite*, and assorted magazines, 1931–1932.

Texas Attorney General, Litigation Files, Archives and Information Services Division, Texas State Library and Archives Commission. Austin.

Red River Bridge Company of Texas v. James W. Good/Dwight F. Davis (Secretary of War of the United States) (#49,307—Equity) (Supreme Court of District of Columbia) (1929–1931).

Red River Bridge Company of Texas et al. v. State of Texas et al. (#50,476) (126th District Court) (1931–1932).

Red River Bridge Company of Texas & Walton Findley, Clyde Findley, E. W. Findley, & Dade Hughes v. State of Oklahoma (District Court, Bryan County, State of Oklahoma) (1929).

Handy, J. R. (Receiver for Red River Bridge Company) et al. v. Cone Johnson et al. (Texas Highway Commission) (#348—Equity) (1931).

Handy, J. R. (Receiver for Red River Bridge Company) et al.—Cone Johnson et al. (State Highway Commission of Texas) v. (5th Circuit Court of Appeals) (1931).

Handy, J. R. (Receiver for Red River Bridge Company) v. Burt McCaleb et al. (#4340—Equity) (1932).

Texas governor Ross S. Sterling. Archives and Information Services Division, Texas State Library and Archives Commission. Austin.

Denison Chamber of Commerce telegram, July 20, 1931.

Denison Resolutions, July 21 and 22, 1931.

J. W. Bailey telegram, July 24, 1931.

Jones letter and clipping, July 18, 1931.

Mills letter, July 22, 1931.

Murray's order opening bridge, July 17, 1931.

Murray's wire to Sterling, July 17, 1931.

Retail Merchants Protective Association of Sherman telegram, July 20, 1931.

Sherman Resolutions, July 21 and 22, 1931.

Tatum letter, July 23, 1931.

Telegram draft, July 18, 1931.

The United States Daily letter, July 23, 1931, and, *The United States Daily* letter, August 5, 1931.

Womack letter, July 23, 1931.

Ross S. Sterling Family Collection, Wallisville Heritage Park, Research Center & Library, Wallisville, TX.

Sterling family members in 1988 donated a sizable collection of items once in the possession of Sterling's late son, Walter Gage Sterling. The collection includes scrapbooks, campaign literature, campaign biographies, photographs, and other materials relating to Sterling's term in office. The collection also includes early family genealogical information and business records. The bulk of the collection consists of records from Sterling's term as chairman of the state highway department, records that might otherwise have been destined for the state archives.

Major flooding in 2005 waterlogged much of the collection. As of mid-2015, only very limited materials are available as staff restores and re-catalogs.

William H. "Alfalfa Bill" Murray Collection, 1891–1962, the Carl Albert Congressional Research and Studies Center, University of Oklahoma, Norman.

Murray to Attorney General King, March 13, 1933.

Rosser to Murray, April 11, 1933.

Rosser to Murray, April 13, 1933.

Text of Murray Speech to 14th Legislature, Special Session, July 13, 1933.

Governors Murray and Sterling

Brief Biography of Murray (Alfalfa Bill): The Statesman and Sage of Oklahoma. Ardmore, OK: McClean Printing, 1929.

Bryant, Keith L. *Alfalfa Bill Murray.* Norman: University of Oklahoma Press, 1968.

Hines, Gordon. *Alfalfa Bill: An Intimate Biography.* Oklahoma City: University of Oklahoma Press, 1932.

Murray, William H. *Memoirs of Governor Murray and True History of Oklahoma: Together with His Biography, Philosophy, Statesmanship, and Oklahoma History, Interwoven.* Boston: Meador, 1945.

Sterling, Ross S., and Edward W. Kilman. *Ross Sterling, Texan: A Memoir by the Founder of Humble Oil & Refining Company*, ed. Don Carleton. Austin: University of Texas Press, 2007.

Keith Bryant's *Alfalfa Bill Murray* is by far the most factual, balanced, and readable of the Murray books. Hines's *Alfalfa Bill* was a campaign biography, hagiographic but interesting as a portrait of Murray as he wanted to be seen in 1931. Murray's own three-volume *Memoirs* is as much family history, recipe book, agricultural treatise, law book, travelogue, and score-settler as a traditional memoir. He barely mentions the Red River Bridge War, and only then as a launchpad for a few folksy stories and a constitutional law lesson. But with every page *Memoirs* offers a peek into the mind of this unique man.

Ross Sterling, Texan was written in the governor's voice from taped interviews. His personal story is an admirable one, but told here with little personality. He makes no mention of the Red River Bridge War and is less than forthcoming about some other missteps during his single term as governor of Texas.

Oklahoma and Texas History

Encyclopedia of Oklahoma History and Culture. Oklahoma City: Oklahoma Historical Society. http://www.okhistory.org/publications/encyclopedia online

Handbook of Texas Online. Texas State Historical Association. https://tshaonline.org/handbook

Calhoun, Sharon C., and Billie J. English. *Oklahoma Heritage.* Maysville, OK: Holt, Calhoun, Clark & Quaid, 1983.

Campbell, Randolph B. *Gone to Texas: A History of the Lone Star State.* New York: Oxford University Press, 2003.

Fischer, LeRoy Henry. *Oklahoma's Governors, 1929–1955: Depression to Prosperity.* Oklahoma City: Oklahoma Historical Society, 1983.

Haley, James L. *Texas: From Spindletop through World War II.* New York: St. Martin's, 1993.

Lovegrove, Michael W. *A Nation in Transition: Douglas Henry Johnston and the Chickasaws, 1898–1939.* Ada, OK: Chickasaw, 2009.

McReynolds, Edwin C. *Oklahoma: A History of the Sooner State.* Norman: University of Oklahoma Press, 1977.

Phares, Ross. *The Governors of Texas.* Gretna, LA: Pelican, 1998.

Gone to Texas is an engaging, entertaining read, even for someone familiar with the outlines of Texas history. *Oklahoma: A History of the Sooner State* is somewhat dated, but it provides a sound introduction to the state's history between 1850 and 1940. The online resources *Encyclopedia of Oklahoma History & Culture* and *Handbook of Texas Online* are great starting points for research. Most entries include a bibliography to facilitate further study.

Highway Department Histories

Biennial Report of Oklahoma Department of Highways. Oklahoma City: Oklahoma Dept. of Highways, 1914–1932.

Burke, Bob. *ODOT 100: Celebrating the First 100 Years of Transportation in Oklahoma.* Oklahoma City: Oklahoma Heritage Association, 2011.

Greer, D. C. *History of the Texas Highway Department.* Rev. ed. Austin: Texas Highway Dept., Information & Statistics Division, 1959.

Oklahoma Highways and History. Oklahoma City: Oklahoma Department of Highways, n.d.

State Highway Department of Texas—Biennial Report. Austin: State Highway Dept. of Texas, 1918–1932.

Read serially, the biennial reports submitted by the Oklahoma and Texas highway departments tell a riveting story of how quickly the states mobilized to build a highway transportation network. Each report includes data on miles built, money spent, road usage, specifications, and much, much more.

Bridge War Narratives

The story of the Red River Bridge War has never been told at length; it is usually tossed off as the same five-hundred-word featurette while repeating the same errors (or straying even further from facts). However, three individuals have produced somewhat longer versions of the story. These narratives were based on good scholarship and independent research, and I referred to them most often as I completed my own narrative.

Fossey, W. Richard. “The Red River Bridge Conflict: A Minor Skirmish in the War against Depression.” *Red River Valley Historical Review* 1, no. 3 (1974): 233.

Lincecum, Jerry B. “Red River Bridge War.” In *Folklore in Motion: Texas Travel Lore,* edited by Kenneth L. Untiedt. Denton: University of North Texas Press, 2007.

Perez, Ruth Ann Overbeck. "The Red River Bridges: A Study in State-Federal Governmental Problems." Master's thesis, University of Texas at Austin. 1969.

Newspapers

Austin American—Austin
Daily Ardmoreite—Ardmore, OK
Daily Oklahoman—Oklahoma City
Dallas Morning News—Dallas
Dallas Times Herald—Dallas
Denison Herald—Denison, TX
Durant Weekly—Durant, OK
Fort Worth Star-Telegram—Fort Worth
Sherman Democrat—Sherman, TX

These newspapers recorded and published events of the bridge war as they occurred. I quote from and cite these newspapers most often because they had reporters making firsthand observations at the scene of events.

Transportation

Borneman, Walter R. *Rival Rails: The Race to Build America's Greatest Transcontinental Railroad.* New York: Random House, 2010.

Brown, Jeffrey R. "The Numbers Game: The Politics of the Federal Surface Transportation Program." PhD diss., University of California, 2003.

Clymer, Floyd. *Those Wonderful Old Automobiles.* New York: McGraw-Hill, 1953.

Davis, Charles Henry, and Stanley E. Bates. *National Highways to Bring about Good Roads Everywhere.* Washington, DC: National Highways Assn., 1913.

Erlichman, Howard J. *Camino Del Norte: How a Series of Watering Holes, Fords, and Dirt Trails Evolved into Interstate 35 in Texas.* College Station: Texas A&M University Press, 2006.

Federal Legislation and Regulations Relating to the Improvement of Federal-Aid Roads and National-Forest Roads and Trails. Washington, DC: US Dept. of Agriculture, US Government Printing Office, 1928.

Highways of Texas. Houston: Humble Oil and Refining, 1927. This map booklet describes routes of the main named highways by milepost or landmark.

Kelly, Susan Croce. *Father of Route 66: The Story of Cy Avery.* Norman: University of Oklahoma Press, 2014.

Klein, Daniel B., and John Majewski. *Turnpikes and Toll Roads in Nineteenth-Century America.* In EH.Net Encyclopedia, ed. Robert Whaples. February 10, 2008. Web.

Logan, Thomas F. "Hastening the Millennium." *Saturday Evening Post,* March 4, 1911, 26–29.

Maguire, Jack. *Katy's Baby: The Story of Denison, Texas.* Austin: Nortex, 1991.

Potter, John E. *The United States Postal Service: An American History.* Washington, DC: Government Relations, USPS, 2007.

Borneman's *Rival Rails* describes in detail the value of rail lines to the towns that acquired them (and the consequences for towns that didn't). It was a lesson not overlooked later by local civic groups as they competed to have interstate named and numbered highways routed through their towns.

Erlichman, in *Camino del Norte,* speculates that the Red River Bridge War may have cost Sherman, Denison, and Durant interstate highway status. Highway 77—not Highway 75—became the general route for I-35 across the Red River.

Roads

Official Good Roads Yearbook of the United States. Washington, DC: American Highway Association, 1914.

Official Red Book and Guide. Dallas: Dallas Automobile Club, 1912.

Ormsby, Waterman Lilly, and Lyle Henry Wright. *The Butterfield Overland Mail.* San Marino, CA: Huntington Library, 1942.

Overbeck, Ruth Ann. "Texas Gets Out of the Mud." *East Texas Historical Journal* 12, no. 1 (1974).

Potter, Isaac B. *The Gospel of Good Roads: A Letter to the American Farmer.* 2d ed. New York: League of American Wheelmen, 1891.

Swift, Earl. *The Big Roads: The Untold Story of the Engineers, Visionaries, and Trailblazers Who Created the American Superhighways.* Boston: Houghton Mifflin Harcourt, 2011.

Wallace, Karl Edward. "Texas and the Good Roads Movement: 1895 to 1948." Master's thesis. University of Texas at Arlington, 2008.

The Great Depression

"Americans in Depression and War." In *The US Department of Labor: Bicentennial History of the American Worker*, edited by Richard B. Morris, 26–31. Washington, DC: US Department of Labor, 1976.

Bernstein, Irving. *The Lean Years: A History of the American Worker*. Baltimore: Penguin, 1970. See pages 421–423, particularly, about the frequency of food riots and looting during the first years of the Great Depression.

Bransten, Richard, and John Stuart. *The Fat Years and the Lean*. New York: Modern Age, 1940.

Carnes, Mark C., and John A. Garraty. *American Destiny: Narrative of a Nation*. New York: Longman, 2003.

Guinn, Jeff. *Go Down Together: The True, Untold Story of Bonnie and Clyde*. New York: Simon & Schuster, 2009.

Himmelberg, Robert F. *The Great Depression and the New Deal*. Westport, CT: Greenwood, 2001.

Ingram, Dale, and Max Brantley. "The Forgotten Rebellion." *Arkansas Times*, January 19, 2006.

Park, David. "Grayson County, Texas, in Depression and War: 1929–1946." Master's thesis, University of North Texas, 2009.

Watkins, T. H. *The Hungry Years: A Narrative History of the Great Depression in America*. New York: Henry Holt, 1999.

Wecter, Dixon. *The Age of the Great Depression, 1929–1941*. New York: Macmillan, 1948.

"Americans in Depression and War," by the US Department of Labor, describes the Depression with numbers, charts, and spreadsheets. The Park thesis localizes that data to Grayson County. *Go Down Together*, *The Hungry Years*, and *The Great Depression and the New Deal* describe the heartbreaking human toll of financial collapse in the United States.

Tolling

Baxandall, Phineas. *Private Roads, Public Costs—the Facts about Toll Road Privatization and How to Protect the Public.* Boston: US PIRG, 2009.

"Business: Toll Bridges." *Time*, April 23, 1928.

Engel, Eduardo, Ronald Fischer, and Alexander Galetovic. "Privatizing Highways in the United States." *Review of Industrial Organization* 29, no. 1 (2006): 27–53.

Ibanez, Jose, and John Robert Meyer. *Going Private: The International Experience With Transport Privatization.* Washington, DC: Brookings Institution, 1993.

"Revival of Turnpike Companies." *Southern Good Roads*, January 1, 1910, 8.

Samuel, Peter. "Texas Has a Rich History of Private Toll Concessions." *TollRoadsNews*, April 19, 2007.

Spivak, Jeffrey. "Toll Roads: The Route to Redevelopment?" *Urban Land: Magazine of the Urban Land Institute*, October 29, 2013.

United States. House of Representatives. *Committee on Interstate and Foreign Commerce.* 71st Cong. Washington, DC: Government Printing Office, 1930. The hearing exhibit on p67 is a chart of revenues, expenses, and net income for the Red River Bridge Company 1890 to 1929.

Toll Roads and Free Roads—Message to the President of the United States Transmitting a Letter from the Secretary of Agriculture, Concurred in by the Secretary of War, Enclosing a Report of the Bureau of Public Roads, United States Department of Agriculture. Washington, DC: Government Printing Office, 1939.

In 1939, with *Toll Roads and Free Roads*, the Bureau of Public Roads determined that interstate toll roads and bridges could not support their own cost for construction and maintenance. Seventy-five years later, private companies are urging state governments to recalculate.

Other Useful Resources

Acheson, Sam Hanna. *Joe Bailey, the Last Democrat.* New York: Macmillan, 1932.

American Motorist. Falls Church, VA: American Automobile Association, 1906–1930.

Annual Report of the Governor of Oklahoma to the Secretary of the Interior, 1904. Washington: Government Printing Office, 1904.

"Bailey, Joseph Weldon—Biographical Information." Biographical Directory of the United States Congress. http://bioguide.congress.gov/scripts/biodisplay.pl?index=B000044.

Benke, Arthur C., and Colbert E. Cushing. *Rivers of North America*. Amsterdam: Elsevier/Academic, 2005.

Braden, G. B. "The Colberts and the Chickasaw Nation." *Tennessee Historical Quarterly* (Winter 1959): 222–41.

Brien, P. J. *Will Rogers, Ambassador of Good Will, Prince of Wit and Wisdom*. Philadelphia: John C. Winston, 1935.

Brown, Norman D. *Hood, Bonnet, and Little Brown Jug: Texas Politics, 1921–1928*. College Station: Texas A&M University Press, 1984.

Caro, Robert A. *The Years of Lyndon Johnson*. Vol. 1, *The Path to Power*. New York: Vintage, 1990.

Census of Manufactures: 1905. Washington, DC: US Government Printing Office, 1906.

Cerami, Charles A. *Jefferson's Great Gamble*. Naperville, IL: Source, 2004.

Chapman, Betty T. "Diverse Legacy of Ross Still Evident All over Houston." *Houston Business Journal*, September 14, 2001.

Cordell, J. William. *Directory of the State of Oklahoma, 1931; State, Congressional and County Officers; Legislatures, First to Thirteenth; Elections, 1907 to 1930; Initiated and Referred Measures*. Wewoka, OK: Wewoka Company, 1931.

Cox, Mike. *The Texas Rangers*. New York: Forge, 2008.

———. *Time of the Rangers: Texas Rangers from 1900 to the Present*. New York: Forge, 2009.

Cranfill, Leslie W. "The Early History of Denison, Texas." Master's thesis, Hardin-Simmons University, 1951.

Crawford, John H., ed. *Red River Bridge War*. Denison, TX: Grayson County Frontier Village, n.d.

Dale, Edward Everett, and Morris L. Wardell. *History of Oklahoma*. New York: Prentice-Hall, 1948.

Dary, David. *Stories of Old-Time Oklahoma*. Norman: University of Oklahoma Press, 2011.

Downing, Jim. "Rugged Road to Canaan." *Oklahoma Today* 1, no. 2 (Spring 1978): 2–7.

Drago, Harry Sinclair. *Red River Valley: The Mainstream of Frontier History from the Louisiana Bayous to the Texas Panhandle.* New York: C. N. Potter, 1962.

Enright, Dennis J. *The Public versus Private Toll Road Choice in the United States.* Hoboken, NJ: NW Financial Group, 2007.

Farrar, R. M. *The Story of Buffalo Bayou and the Houston Ship Channel.* Houston: Chamber of Commerce, 1926.

Federal Road-Aid Act (also called Federal Aid Road Act). Circ. No. 109. Washington, DC: US Department of Agriculture, 1928.

"Finish of Fergusonism." *Time,* September 1, 1930.

Fleming, Thomas. *The Louisiana Purchase.* New York: Wiley, 2003.

Foreman, Grant, "Red River and the Spanish Boundary in the Supreme Court," *Chronicles of Oklahoma* 2 (March 1924): 298–310.

———. "The California Overland Mail Route through Oklahoma." *Chronicles of Oklahoma* 9, no. 3 (September 1931): 300–313.

———. *Down the Texas Road: Historic Places along Highway 69 through Oklahoma.* Norman: University of Oklahoma Press, 1936.

Gammel, Hans Peter Mareus Neilsen. *The Laws of Texas 1822–1897.* Vols. 7 and 28. Austin: Gammel Book Publishing, 1898.

Green, Richard. *Chickasaw Lives.* Vol. 1–4. Ada, Oklahoma: Chickasaw Press, 2007.

Greene, A. C. *900 Miles on the Butterfield Trail.* Denton: University of North Texas Press, 1994.

Guttery, Ben R. *Representing Texas: A Comprehensive History of US and Confederate Senators and Representatives from Texas.* Charleston, SC: BookSurge, 2008.

Hays, Robert E. "Old School Journalist: A Biography of Claud Easterly." Commerce, TX: Master's thesis, East Texas State University, 1972.

Henry, Robert. "The Sage of Tishomingo," *Oklahoma Today* 57 (2007): 5.

Herman, Alan. "Dust, Depression, and Demagogues: Political Radicals of the Great Plains 1930–1936." *Journal of the West* 16, no. 1 (1977): 57–62.

Hillerman, Tony. *Seldom Disappointed: A Memoir.* New York: HarperCollins, 2001.

Hunt, Donna. "Judge Jake Loy." *Texoma Living,* no. 3 (June 1, 2007).

"Joseph Weldon Bailey, Jr. '15." *Princeton Alumni Weekly,* October 15, 1943, 9.

Karrar, Ray E. "Invention in the Congressional and Campaign Speaking of Joe Weldon Bailey." Master's thesis. North Texas State University, 1968.

Kasson, John F. *The Little Girl Who Fought the Great Depression: Shirley Temple and 1930s America.* New York: W. W. Norton, 2014.

Kelso, Harold. "'Navigable Waters' as a Legal Fiction." *Journal of Land & Public Utility Economics* 17, no. 4 (November 1941): 394–405.

Kent, Noel J. *America in 1900.* Armonk, NY: M. E. Sharpe, 2002.

Kukla, Jon. *A Wilderness So Immense.* New York: Random House, 2003.

Larson, Henrietta M., and Kenneth Wiggins Porter. *History of Humble Oil & Refining Company: A Study in Industrial Growth.* New York: Harper, 1959.

Lewis, James E. *John Quincy Adams: Policymaker for the Union.* Wilmington, DE: Rowman & Littlefield, 2001.

Liles, Maurine Walpole. *Sam and the Speaker's Chair: The Story of Sam Rayburn, Speaker of the US House of Representatives.* Austin: Eakin, 1994.

Lincecum, Jerry Bryan, editor. *Telling Our Stories: Grayson County Reminiscences: The First 150 Years, 1846–1996.* Sherman, TX: Big Barn, 1996.

Lovegrove, Michael W. *A Nation in Transition: Douglas Henry Johnston and the Chickasaws, 1898–1939.* Ada, OK: Chickasaw, 2009.

Luthin, Reinhard H. *American Demagogues: Twentieth Century.* Gloucester, MA: Peter Smith, 1959.

Mallory, John A. *United States Compiled Statutes, Annotated, 1916: Embracing the Statutes of the United States of a General and Permanent Nature in Force at the Close of the First Session of the 64th Congress and Decisions Construing and Applying Same to April 1, 1916,* Vol. 10. St. Paul: West, 1916.

Marshall, Thomas Maitland. *A History of the Western Boundary of the Louisiana Purchase, 1819–1841.* Vol. 2. Berkeley: University of California Press, 1914.

McKay, Seth Shepard, and Odie B. Faulk. *Texas after Spindletop, 1901–1965.* Austin: Steck-Vaughn, 1965.

McKay, Seth Shepard. *W. Lee O'Daniel and Texas Politics, 1938–1942,* Lubbock: Texas Technological College Research Funds, 1944.

McKellips, Karen, "Inside the School Yard Gate: 'Alfalfa Bill' Murray and Education in Oklahoma," *Chronicles of Oklahoma* 78 (January 2001).

McNichol, Dan. *The Roads That Built America: The Incredible Story of the US Interstate System.* New York: Sterling, 2006.

Milburn, George. "The Sage of Tishomingo." *The American Mercury* 23, no. 89 (May 1931): 11–19.

Modern Highway, Jefferson Highway Association, 5: (January–December 1920).

Morehouse, Ward. *Just the Other Day: From Yellow Pines to Broadway*. New York: McGraw-Hill, 1953.

Morrison, W. B. (recorder). "Colbert Ferry on Red River, Chickasaw Nation, Indian Territory: Recollections of John Malcolm, Pioneer Ferryman," *Chronicles of Oklahoma* 16, no. 3 (September 1938).

Nickels, Lenora. "Public Services of Dan Moody." Master's thesis, Texas Tech University, 1948

Nieman, Robert. "Twentieth Century Shining Star: William Warren Sterling" *Texas Ranger Dispatch*, January 1, 2004, 8–9.

"Notes and Documents." *Chronicles of Oklahoma*, 1957, 91–109.

Overacker, Louise, and Victor J. West. *Money in Elections*. New York: Macmillan, 1932.

Poole, Robert, and Samuel Peter. "The Return of Private Toll Roads." *Public Roads*, March 1, 2006.

Procter, Ben H. *Just One Riot: Episodes of Texas Rangers in the 20th Century*. Austin: Eakin, 2000.

Samuel, Peter, and Robert W. Poole. *Should States Sell Their Toll Roads?* Los Angeles: Reason Public Policy Institute, 2005.

Schlesinger, Arthur M. *The Crisis of the Old Order, 1919–1933*. Boston: Houghton Mifflin, 1957.

Schraff, Anne E., and Barbara Silberdick Feinberg. *The Great Depression and the New Deal: America's Economic Collapse and Recovery*. New York: F. Watts, 1990.

Schruber, Francis W. "The Return of 'Alfalfa Bill' Murray" *Chronicles of Oklahoma* 41 (March 1963).

Shepherd, William G. "King of the Prairie." *Collier's*, November 28, 1931, 12–16.

Smith, Corinne. *Historic American Engineering Record: Red River Bridge at US Highway 82, Garland, Miller County, AR*. Washington, DC: National Park Service, Department of the Interior, 1988.

Southerland, Edward. "Building Denison Dam." *Texoma Living!*, August 1, 2010.

Souvenir of Denison, the Gate City of Texas. Denison, TX: News Steam Printing House, 1887.

Starr, Kevin. *Golden Gate: The Life and Times of America's Greatest Bridge*. New York: Bloomsbury, 2010.

"Texas Legislators: Past & Present." *Legislative Reference Library*, August 9, 2014.

Thornton, Emily. "Roads to Riches." *Business Week*, May 7, 2007, 12–19.

Toll Facilities in the United States. Publication No: FHWA-PL-13–037. Washington, DC: Office of Highway Policy Information, Federal Highway Administration, September 2013.

US Department of the Interior. *Laws, Decisions, and Regulations Affecting the Work of the Commissioner to the Five Civilized Tribes 1893 to 1906.* Washington DC: Government Printing Office, 1906.

US House of Representatives. *Red River Bridges. Hearing before the Committee on Interstate and Foreign Commerce, February 10 and 11, 1930.* 71st Cong. Washington, DC: Government Printing Office, 1930.

Wilson, Carol O. *In the Governor's Shadow: The True Story of Ma and Pa Ferguson.* Denton: University of North Texas Press, 2014.

Wright, Murial H. "Historic Places on the Old Stage Line from Fort Smith to Red River." *Chronicles of Oklahoma*, 1931, 798–814.

Wright, Murial H. "The Butterfield Overland Mail One Hundred Year Ago." *Chronicles of Oklahoma*, 1957, 55–71.

Yearbook of the United States Department of Agriculture, 1901–1919. Washington, DC: Government Printing Office.

INDEX